ALGEBRAIC EQUATIONS

ALGEBRAIC EQUATIONS

AN INTRODUCTION TO THE THEORIES OF LAGRANGE AND GALOIS

BY

EDGAR DEHN, PH.D.

Instructor in Mathematics at Columbia University

DOVER PUBLICATIONS, INC.

NEW YORK

Published simultaneously in Canada by McClelland and Stewart, Ltd.

Published in the United Kingdom by Constable and Company Limited, 10 Orange Street, London, W.C. 2.

This new Dover edition, first published in 1960, is an unabridged and corrected republication of the work first published by the Columbia University Press in 1930.

Manufactured in the United States of America

Dover Publications, Inc.
180 Varick Street
New York 14, N.Y.

CONTENTS

LIST OF THEOREMS

LIST OF ADOPTED CONVENTIONS

While it is neither possible nor desirable to stereotype algebraic notation, certain conventions are expedient if waived when inopportune.

$\sum$	sum of functions
$\prod$	product of functions
x	argument of function
x_i, α_i	roots of function
f, g	function of x
φ, ψ	function of the x_i
$\varphi(x_i)_1^n, \varphi(x_i) = \varphi(x_1, \ldots, x_n)$	
ψ_i	conjugate functions
F	reducible function
f	irreducible function
R, r	rational function*
j	integral function
c	cyclotomic function
S	symmetric function* of the x_i
s	symmetric sum* in the x_i
$s^{(n)}$	same on n letters x_i
A	alternating function of the x_i
R	resultant
D	discriminant in homogeneous form
Δ	discriminant in non-homogeneous form
G_2, G_3	invariants of cubic
g_2, g_3	invariants of biquadratic
W	weight of function
D	total degree of function
v	elementary Galoisian function
$g(v)$	Galoisian function
$G(v)$	complete Galoisian function

* Symbols for symmetric and rational functions may be used without reference to a definite form of those functions.

$g(v)$	primary Galoisian function
$g(v) = 0$	Galoisian resolvent
$r(\psi) = 0$	ordinary resolvent
(ϵ,ψ)	Lagrange's solvent
s,t	permutation (substitution) on the x_i
1	identical permutation
$(x_1x_i)_2^n$	$= (x_1x_2), \ldots, (x_1x_n)$
G	group of permutations on the x_i
$\{G\}$	explicit notation of groups
φ	function belonging to G
S	symmetric group
A	alternating group
1	identical group
v	function belonging to 1
H	subgroup of G
ψ	function belonging to H
ξ	function belonging to group that contains H but no other permutation of G
Ξ	group of ξ
t	permutation in G but not in H
H_t	conjugate subgroup
Ht	partition of G
D	greatest common subgroup
N, J	normal (invariant) subgroup
$\overline{N}$	maximum normal subgroup
X_i	group of x_i
Z	central subgroup
z	normal permutation
V	quadratic group
M	metacyclic group
Q	quotient of permutation-groups
q	permutation in Q
$\{s,t\}$	group generated by s and t
$C,\{s\}$	cyclic group $s,s^2, \ldots$
$\{s_i\}_1^n,\{s_i\}$	group $s_1,s_2, \ldots, s_n$
$\langle G\rangle,G$	Galoisian group
$\langle A\rangle,A$	Abelian group
C	class of permutations

c	permutation in C
σ,τ	permutation on the ψ_i
Γ	group of such permutations
σ	substiiution in domain
$\langle\Gamma\rangle$	substitution-group = $\langle G\rangle$
G/N	abstract quotient or factor-group
a,b	element
1	identical element
r,ρ	order of group or permutation
r_g,g	order of group G
G_r	group of order r
n	degree of group or permutation
G^n	group of degree n
r_s	order of permutation s
ρ	relative order of permutation
j	index of group
j_g	index of G in S
j_{hg}	index of H in G
Ω	domain
(a)	$= \Omega(a)$
$(a_i)_1^n,(a_i)$	$= (a_1, \ldots, a_n)$
ω	number in Ω
θ	primitive number in Ω
p	prime number
g	primitive root of p
r	rational number
ϵ	primitive root of unity
ω	primitive cube root of unity
i	primitive fourth root of unity

$\varphi(n)$ number of positive integers prime to n and smaller than n

ALGEBRAIC EQUATIONS

CHAPTER I

INTEGRAL FUNCTION

§1. INTERPOLATION

It is the test of a perfect theory that its study leaves in the mind of the student a sense of accomplishment and beauty. One such theory is that theory of algebraic equations which we owe to two men of genius: Lagrange and Galois.

The solution of equations is the principal object of **algebra,** as the original meaning of this term implies, but it is understood that the means of an **algebraic solution** are confined to **algebraic operations** alone which, beside the rational operations of arithmetic, include the extraction of roots.[1]

It will be noticed that equations in one unknown only are treated here and that the problem of elimination is not under discussion.

An algebraic equation is of the form

$$F(x) = 0$$

where $F(x)$ is an algebraic function. As single-valued algebraic functions are necessarily rational,[2] we may set

$$F(x) = \frac{f(x)}{g(x)}$$

with integral $f(x)$ and $g(x)$ and proceed to an equation

$$f(x) = 0$$

where $f(x)$ is an integral rational function.

[1] By such operations we mean here nothing more than the use of the proper signs, by extraction of roots nothing more than the use of the radical sign.

[2] Cf. Townsend, Functions of a Complex Variable, Art. 54, p. 290.

Hence we begin by noting some properties of integral rational functions,[1] or polynomials. Every such function of the variable x can be represented in the form

$$f(x) = a_0x^n + a_1x^{n-1} + \dots + a_{n-1}x + a_n.$$

So to determine the $n + 1$ coefficients a_i that for $n + 1$ given values

$$x_0, x_1, \dots, x_n$$

of the variable the function assumes $n + 1$ given values

$$y_0, y_1, \dots, y_n$$

is a problem of interpolation. It is solved by the

Interpolation formula of Lagrange:

$$\begin{aligned} f(x) = {} & y_0\frac{(x - x_1)(x - x_2)(x - x_3) \dots (x - x_n)}{(x_0 - x_1)(x_0 - x_2)(x_0 - x_3) \dots (x_0 - x_n)} \\ & + y_1\frac{(x - x_0)(x - x_2)(x - x_3) \dots (x - x_n)}{(x_1 - x_0)(x_1 - x_2)(x_1 - x_3) \dots (x_1 - x_n)} \\ & + y_2\frac{(x - x_0)(x - x_1)(x - x_3) \dots (x - x_n)}{(x_2 - x_0)(x_2 - x_1)(x_2 - x_3) \dots (x_2 - x_n)} \\ & \quad \cdot \quad \cdot \quad \cdot \quad \cdot \quad \cdot \\ & + y_n\frac{(x - x_0)(x - x_1) \dots (x - x_{n-1})}{(x_n - x_0)(x_n - x_1) \dots (x_n - x_{n-1})}. \end{aligned}$$

Setting

$$g(x) = (x - x_0)(x - x_1)(x - x_2) \dots (x - x_n).$$

we have as its derivative

$$\begin{aligned} g'(x) = {} & (x - x_1)(x - x_2) \dots (x - x_n) \\ & + (x - x_0)(x - x_2) \dots (x - x_n) \\ & \quad \cdot \quad \cdot \quad \cdot \quad \cdot \quad \cdot \\ & + (x - x_0)(x - x_1) \dots (x - x_{n-1}), \end{aligned}$$

whence

$$\begin{aligned} g'(x_0) &= (x_0 - x_1)(x_0 - x_2) \dots (x_0 - x_n) \\ g'(x_1) &= (x_1 - x_0)(x_1 - x_2) \dots (x_1 - x_n) \\ & \quad \cdot \quad \cdot \quad \cdot \quad \cdot \quad \cdot \\ g'(x_n) &= (x_n - x_0)(x_n - x_1) \dots (x_n - x_{n-1}). \end{aligned}$$

[1] If no confusion can arise, we shall simply call them integral functions.

Consequently, the interpolation formula can be written

$$f(x) = g(x)\left[y_0 \frac{1}{g'(x_0)(x - x_0)} + y_1 \frac{1}{g'(x_1)(x - x_1)} + \cdots \right]$$

or

$$f(x) = g(x) \sum_{i=0}^{n} \frac{y_i}{g'(x_i)(x - x_i)}.$$

§2. DIVISION

If

$$f = a_0x^n + a_1x^{n-1} + \cdots + a_n$$

and

$$\varphi = \alpha_0x^\nu + \alpha_1x^{\nu-1} + \cdots + \alpha_\nu$$

are two integral functions of x such that

$$n \geqq \nu,$$

we form the integral function

$$f_1 = f - \frac{a_0}{\alpha_0}x^{n-\nu} \cdot \varphi$$

which is of degree smaller than n. Then setting

$$f_1 = b_0x^m + b_1x^{m-1} + \cdots + b_m,$$

we form the integral function

$$f_2 = f_1 - \frac{b_0}{\alpha_0}x^{m-\nu} \cdot \varphi$$

which is of degree smaller than m. So we continue until we reach the integral function

$$f_k = f_{k-1} - \frac{p_0}{\alpha_0}x^{h-\nu} \cdot \varphi$$

which is of degree smaller than ν and then adding obtain

$$f = \frac{a_0x^{n-\nu} + b_0x^{m-\nu} + \cdots + p_0x^{h-\nu}}{\alpha_0}\varphi + f_k$$

or

$$f = q \cdot \varphi + r.$$

Since such an operation is ordinary division of two integral functions f and φ giving q as quotient and r as remainder, we can note that

(1) **quotient and remainder in the division of two integral functions are themselves integral in the variable of those functions,**

and also in their coefficients when $\alpha_0 = 1$:

$$q = j_1(x\,|a_i|\,\alpha_i)$$
$$r = j_2(x\,|a_i|\,\alpha_i).$$

It appears now that **Euclid's algorism** for finding the greatest common factor of two integers can be applied to two integral functions f and f_1 of x. For we have

$$f = q_1 f_1 + f_2$$
$$f_1 = q_2 f_2 + f_3$$
$$\cdots\cdots\cdots$$
$$f_{r-3} = q_{r-2} f_{r-2} + f_{r-1}$$
$$f_{r-2} = q_{r-1} f_{r-1} + f_r,$$

which terminates since the degrees of the f_i constantly decrease until f_r becomes independent of x. If f and f_1 have a common factor other than a constant, this factor is evidently contained in every following f_i and

$$f_r = 0.$$

But then, by the last equation of Euclid's algorism, f_{r-1} is a factor of f_{r-2}, by the preceding equation it is a factor of f_{r-3}, . . . , and hence it is a factor also of f_1 and f. Moreover, it is the greatest common factor of f and f_1 since it contains every common factor of f and f_1, and so it follows that

(2) **the greatest common factor of two integral functions is computable by rational operations.**

Substituting the value of f_2 obtained from the first equation of Euclid's algorism into the second, we find

$$f_3 = (1 + q_1 q_2) f_1 - q_2 f,$$

or abbreviated:

$$f_3 = g_1 f + g f_1.$$

Expressing by similar substitutions the value of every following f_i in terms of f and f_1, we finally obtain

$$f_r = G_1 f + G f_1.$$

In this equation G and G_1 are integral functions of x, but they are not uniquely defined since the equation remains true when we replace G by $G + Hf$ and G_1 by $G_1 - Hf_1$, where H is any integral function of x. There is only one function G_1, however, of degree not more than $n_1 - 1$, so that $f_r - G_1 f$ is of degree not more than $n + n_1 - 1$, if n and n_1 are the degrees of f and f_1; and corresponding to this only one function G of degree not more than $n - 1$.

If f and f_1 have no common factor, f_r is a constant different from zero, and by dividing it out we prove:

(3) **If two integral functions f and f_1 of degree n and n_1 have no common factor, we can compute by rational operations two integral functions φ and φ_1 which satisfy identically the relation**

$$\boxed{\varphi_1 . f + \varphi . f_1 = 1}$$

and are uniquely defined of degree not more than $n - 1$ and $n_1 - 1$, if we so choose.

Multiplying this equation by any integral function g of x, we have

$$g\varphi_1 f + g\varphi f_1 = g,$$

and setting

$$g\varphi = qf + r,$$

where q and r are integral functions of x and r is of degree smaller than n, we have

$$g\varphi_1 f + qff_1 + rf_1 = g,$$

where we may put:

$$r_1 = g\varphi_1 + qf_1.$$

From this we conclude:

(4) **If f and f_1 and g are integral functions of x such that f and f_1 have no common factor, we can compute by rational operations two integral functions r and r_1 of x which satisfy identically the relation**

$$\boxed{r_1 . f + r . f_1 = g},$$

and r such that it is of lower degree than f.

§3. REDUCTION

An integral function of x is said to be **reducible** if it is expressible as product of some integral functions of x other than constants and **irreducible** if it is not expressible so.

The following proposition is quickly verified:

(5) **If an integral function of x divides the product of two other integral functions of x, then, having no factor in common with one of them, it divides the other.**

For if the integral function f of x divides the product of two integral functions g and h of x and has no factor in common with g, we can find by proposition (3) two integral functions φ and ψ of x such that

$$\varphi f + \psi g = 1,$$

and from

$$\varphi f h + \psi g h = h$$

follows that f divides h since it divides both terms on the left.

The proposition is readily extended to include the product of more than two functions and as such leads to another:

(6) **A reducible function of x is expressible as product of irreducible factors in only one way, if we disregard constants.**

For suppose the reducible function F of x is expressible so in two ways:

$$F = f_1 f_2 \ . \ . \ . = g_1 g_2 \ . \ . \ .$$

Then the irreducible function g_1 must divide some function f_i, say f_1, and differ from it by a constant factor alone since f_1 also is irreducible:

$$f_1 = k_1 g_1.$$

From

$$k_1 f_2 \ . \ . \ . = g_2 \ . \ . \ .$$

follows likewise that g_2 differs from f_2, say, by a constant factor alone:

$$f_2 = k_2 g_2,$$

and so on. This proves the proposition.

§4. PRIMITIVE FUNCTION

If an integral function has integral coefficients, the greatest common factor of those coefficients is called the **divisor of the function,** and if the divisor is 1, the function is said to be **primitive.**

Let

$$g = g_0x^m + g_1x^{m-1} + \dots + g_m$$

and

$$h = h_0x^n + h_1x^{n-1} + \dots + h_n$$

be two primitive functions and

$$F = a_0x^{m+n} + a_1x^{m+n-1} + \dots + a_{m+n}$$

their product. Suppose that the coefficients

$$g_0, g_1, \dots, g_{k-1}$$
$$h_0, h_1, \dots, h_{l-1}$$

are divisible by some prime number p, while the coefficients g_k and h_l are not. Then

$$a_{k+l} = g_kh_l + g_{k-1}h_{l+1} + \dots$$
$$+ g_{k+1}h_{l-1} + \dots$$

is not divisible by p either, inasmuch as all its terms except the first one are divisible. Hence F is primitive and it appears that

(7) **the product of primitive functions is primitive.**

Since the product of the imprimitive functions pg and qh, where p and q are integers, evidently is divisible by pq, we may add that

(8) **the divisor of a product of imprimitive functions is the product of the divisors of those functions.**

If an integral function of the form

$$x^n + a_1x^{n-1} + \dots + a_n$$

has fractional coefficients, we can represent it as fraction of a primitive function bringing its terms over the lowest common denominator. It follows by proposition (7) that the product of two such functions is of the same form and hence also has fractional coefficients.

To recognize a primitive function

$$f = x^n + a_1x^{n-1} + \dots + a_n,$$

where x^n has the coefficient

$$a_0 = 1,$$

as irreducible, the following proposition[1] is helpful:

(9) If the coefficients a_i other than a_0 of a primitive function f are divisible by a prime number p while $a_0 = 1$ and a_n is not divisible by p^2, the function f is irreducible.

For suppose that

$$f = gh,$$

where

$$g = x^k + g_1x^{k-1} + \ldots + g_k$$
$$h = x^l + h_1x^{l-1} + \ldots + h_l.$$

Then g and h have integral coefficients since f has.

Comparing coefficients, we set

$$\begin{aligned} a_n &= g_kh_l \\ a_{n-1} &= g_kh_{l-1} + g_{k-1}h_l \\ a_{n-2} &= g_kh_{l-2} + g_{k-1}h_{l-1} + g_{k-2}h_l \\ &\cdot\quad\cdot\quad\cdot\quad\cdot\quad\cdot\quad\cdot\quad\cdot \\ a_{n-k} &= g_kh_{l-k} + g_{k-1}h_{l-k+1} + \ldots + g_1h_{l-1} + h_l, \end{aligned}$$

where the h_i with possible negative subscripts are 0 and where

$$h_0 = 1.$$

If now every a_i other than a_0 is divisible by p and a_n not by p^2, we infer from the first equation that either g_k or h_l is divisible by p and the other is not. But when g_k is divisible by p, we infer from the following equations that $g_{k-1}, g_{k-2}, \ldots, g_1$ are divisible by p, and from the last equation that h_l is divisible so, which is a contradiction.

As the same argument applies when h_l is divisible by p, the proposition is proved.

§5. LINEAR FACTORS

Dividing an integral function

$$f(x) = a_0x^n + a_1x^{n-1} + \ldots + a_{n-1}x + a_n$$

by the linear function $x - \alpha$, we obtain

$$f(x) = (x - \alpha)g(x) + c,$$

[1] Called the proposition of Eisenstein.

where the remainder c is a constant since as function of x it is of lower degree than $x - \alpha$. Substituting from

$$g(x) = b_0x^{n-1} + b_1x^{n-2} + \ldots + b_{n-2}x + b_{n-1}$$

with degree one less than the degree of $f(x)$, we have

$$\begin{aligned} f(x) = b_0x^n + b_1x^{n-1} &+ \ldots + b_{n-1}x + c \\ - \alpha b_0x^{n-1} &- \ldots - \alpha b_{n-2}x - \alpha b_{n-1}, \end{aligned}$$

whence

$$\begin{aligned} b_0 \quad &= a_0 \\ b_1 - \alpha b_0 &= a_1 \\ \cdot \quad \cdot \quad \cdot &\quad \cdot \quad \cdot \quad \cdot \\ b_{n-1} - \alpha b_{n-2} &= a_{n-1} \\ c - \alpha b_{n-1} &= a_n \end{aligned}$$

and consequently

$$\begin{aligned} b_0 &= a_0 \\ b_1 &= a_0\alpha + a_1 \\ &\cdot \quad \cdot \quad \cdot \quad \cdot \quad \cdot \\ b_{n-1} &= a_0\alpha^{n-1} + a_1\alpha^{n-2} + \ldots + a_{n-1} \\ c &= a_0\alpha^n + a_1\alpha^{n-1} + \ldots + a_{n-1}\alpha + a_n. \end{aligned}$$

Comparing now $f(x)$ with c, we find that

$$c = f(\alpha);$$

hence

$$g(x) = \frac{f(x) - f(\alpha)}{x - \alpha},$$

which for $x = \alpha$ gives

$$g(\alpha) = f'(\alpha)$$

if we assume here the continuity of integral functions[1] and introduce the derivative $f'(x)$ of $f(x)$.

This clears the way for the discussion of roots. When

$$f(\alpha) = 0,$$

we call α a **root** of the function $f(x)$ and have from the preceding

$$f(x) = (x - \alpha)g(x).$$

[1] The continuity of integral functions underlies also the fundamental theorem of algebra which is used on the next page. For the continuity of integral functions cf. Townsend, Functions of a Complex Variable, Ch. II, and Burnside-Panton, Theory of Equations, Art. 7, p. 9, and Art. 192, p. 427.

By the **fundamental theorem of algebra,**[1] every integral function has a root, whence it follows as before that

$$g(x) = (x - \beta)g_1(x)$$
$$\cdot \quad \cdot \quad \cdot \quad \cdot \quad \cdot \quad \cdot$$
$$g_{n-2}(x) = a_0(x - \nu);$$

consequently

$$f(x) = a_0(x - \alpha)(x - \beta) \ . \ . \ . \ (x - \nu).$$

It appears that

(10) **an integral function of degree n is the product of n linear factors,**

and that

(11) **an integral function of degree n has just n roots.**

It could not have more roots unless it would vanish identically with coefficients equal to zero, for the degrees of $g, g_1, \ . \ . \ .$ constantly decrease by one.

Since we have

$$g(\alpha) = a_0(\alpha - \beta) \ . \ . \ . \ (\alpha - \nu) = f'(\alpha),$$

it follows that

(12) **the function $f(x)$ has a multiple root α when**

$$\boxed{f(\alpha) = f'(\alpha) = 0},$$

that is to say when not only the function but also its derivative vanishes for α.

[1] See footnote on preceding page. For the fundamental theorem of algebra cf. Townsend, Functions of a Complex Variable, Art. 54, p. 291, and Burnside-Panton, Theory of Equations, Art. 195, p. 431; see also the end of §82.

CHAPTER II

EQUATIONS AND PERMUTATIONS

§6. DISCOVERY OF LAGRANGE

Easy as the solution of the linear equation is, and versed as we are in the solution of the quadratic, the solution of equations of higher degrees becomes increasingly difficult if not impossible. And it is the study of these equations which gave to us the celebrated theories of Lagrange and Galois and with them the first investigations into **groups** and the first dim notion of **domains**: two concepts that were destined to predominate in modern algebra.

We consider a discovery of Lagrange. Given the general cubic equation

$$x^3 - bx^2 + cx - d = 0$$

whose roots we denote by x_i, we set

$$x = y + \frac{b}{3}$$

and obtain the reduced cubic equation

$$y^3 + py - q = 0,$$

where

$$p = c - \frac{b^2}{3}$$

and

$$q = d - \frac{bc}{3} + \frac{2b^3}{27}.$$

Its roots are by **Cardan's formula**

$$y_1 = \sqrt[3]{\frac{q}{2} + \sqrt{R}} + \sqrt[3]{\frac{q}{2} - \sqrt{R}}$$

$$y_2 = \omega^2 \sqrt[3]{\frac{q}{2} + \sqrt{R}} + \omega \sqrt[3]{\frac{q}{2} - \sqrt{R}}$$

$$y_3 = \omega \sqrt[3]{\frac{q}{2} + \sqrt{R}} + \omega^2 \sqrt[3]{\frac{q}{2} - \sqrt{R}},$$

where

$$R = \frac{q^2}{4} + \frac{p^3}{27},$$

and where

$$\omega = \frac{-1 + \sqrt{-3}}{2}, \quad \omega^2 = \frac{-1 - \sqrt{-3}}{2}.$$

are primitive cube roots[1] of unity for which

$$\omega^2 + \omega + 1 = 0,$$
$$\omega^3 = 1.$$

To find x_i, we only have to add $b/3$ to y_i; whence in the expressions for x_i there will be no radicals save those contained in the expressions for y_i. And these radicals Lagrange noticed to be expressible as functions of the roots x_i themselves.

To see this for the cube roots of the formula, we multiply the equations for the y_i by 1, ω, ω^2 respectively and adding them obtain

$$3\sqrt[3]{\frac{q}{2} + \sqrt{R}} = y_1 + \omega y_2 + \omega^2 y_3.$$

Substituting then for $y_i = x_i - \frac{b}{3}$, we have

$$3\sqrt[3]{\frac{q}{2} + \sqrt{R}} = x_1 + \omega x_2 + \omega^2 x_3.$$

This equation has a cube root as member. Setting

$$\varphi_1 = x_1 + \omega x_2 + \omega^2 x_3 = 3\sqrt[3]{\frac{q}{2} + \sqrt{R}},$$

we know from elementary algebra that the other two cube roots are

$$\varphi_3 = \omega^2 \varphi_1 = 3\omega^2 \sqrt[3]{\frac{q}{2} + \sqrt{R}}$$

and

$$\varphi_5 = \omega \varphi_1 = 3\omega \sqrt[3]{\frac{q}{2} + \sqrt{R}}.$$

Again, multiplying the equations for the y_i by 1, ω^2, ω respectively, adding them and substituting for y_i, we obtain

$$3\sqrt[3]{\frac{q}{2} - \sqrt{R}} = x_1 + \omega^2 x_2 + \omega x_3.$$

[1] Primitive roots are defined in §§38, 79 and computed in §84.

And if we set

$$\varphi_2 = x_1 + \omega^2 x_2 + \omega x_3 = 3\sqrt[3]{\frac{q}{2} - \sqrt{R}},$$

the other two cube roots are

$$\varphi_4 = \omega\varphi_2 = 3\omega\sqrt[3]{\frac{q}{2} - \sqrt{R}},$$

and

$$\varphi_6 = \omega^2\varphi_2 = 3\omega^2\sqrt[3]{\frac{q}{2} - \sqrt{R}}.$$

To find $\sqrt{R}$ in terms of the roots x_i, we cube the equations for φ_1 and φ_2 and subtract. We obtain

$$54\sqrt{R} = (x_1 + \omega x_2 + \omega^2 x_3)^3 - (x_1 + \omega^2 x_2 + \omega x_3)^3,$$

and after expanding, simplifying and factoring we have

$$18\sqrt{R} = \sqrt{-3}\,(x_1 - x_2)(x_1 - x_3)(x_2 - x_3),$$

where

$$\sqrt{-3} = \omega - \omega^2.$$

The other square root is $-\sqrt{R}$.

§7. SOLUTION OF CUBIC

The solution of the general cubic equation requires the computation of the six values φ_i, but is then readily found from the equations

$$\varphi_1 = x_1 + \omega x_2 + \omega^2 x_3$$
$$\varphi_2 = x_1 + \omega^2 x_2 + \omega x_3$$
$$b = x_1 + x_2 + x_3$$

if we add them as they stand, and multiplied by ω^2, ω, 1, and then by ω, ω^2, 1 respectively. It may be found also from two other values φ_i if we take care to choose them so that their product is equal to

$$3\sqrt[3]{\frac{q}{2} + \sqrt{R}} \cdot 3\sqrt[3]{\frac{q}{2} - \sqrt{R}} = -3p,$$

which is the product of φ_1 and φ_2.

We can compute the six values φ_i when we know the two values of their cubes. Setting

$$A_1 = \frac{q}{2} + \sqrt{R}$$

and

$$A_2 = \frac{q}{2} - \sqrt{R},$$

we have the equations

$$\left(\frac{\varphi_j}{3}\right)^3 = A_1 \qquad [j = 1, 3, 5$$

and

$$\left(\frac{\varphi_k}{3}\right)^3 = A_2 \qquad [k = 2, 4, 6$$

to find the φ_i. The A_i we then recognize as the roots of the quadratic equation

$$A^2 - qA - \frac{p^3}{27} = 0.$$

This last equation cannot be avoided either for the six values φ_i satisfy the equation

$$(\varphi - \varphi_1)(\varphi - \varphi_2)(\varphi - \varphi_3)(\varphi - \varphi_4)(\varphi - \varphi_5)(\varphi - \varphi_6) = 0,$$

and multiplying out we obtain the quadratic equation

$$\varphi^6 - (\varphi_1^3 + \varphi_2^3)\varphi^3 + \varphi_1^3\varphi_2^3 = 0$$

in φ^3 which is reducible[1] to the other in A.

The coefficients q and $p^3/27$ which permit us to compute the two values A_i and the six values φ_i have just one value as they are rational functions of the coefficients b, c, d.

Thus, retracing the solution of the general cubic equation, we begin with its coefficients of one definite value. We solve a rational quadratic equation and find the two values $A_i = q/2 \pm \sqrt{R}$, introducing or **adjoining** the radical $\sqrt{R}$ not contained in the coefficients. Then we find the six values φ_i extracting the cube roots of A_i, and adjoining them we compute the roots x_i of the cubic.

[1] Cf. Serret's Algebra No. 508.

§8. CONNECTION WITH PERMUTATIONS

All the quantities that we operate with in solving the general cubic equation are expressible as functions of the roots x_i themselves of the cubic.

Tabulating all φ_i as functions of the x_i and rearranging the order of their terms, we have

$$\begin{aligned}
\varphi_1 &= x_1 + \omega x_2 + \omega^2 x_3 \\
\varphi_3 &= x_2 + \omega x_3 + \omega^2 x_1 = \omega^2 \varphi_1 \\
\varphi_5 &= x_3 + \omega x_1 + \omega^2 x_2 = \omega \varphi_1 \\
\varphi_2 &= x_1 + \omega x_3 + \omega^2 x_2 \\
\varphi_4 &= x_2 + \omega x_1 + \omega^2 x_3 = \omega \varphi_2 \\
\varphi_6 &= x_3 + \omega x_2 + \omega^2 x_1 = \omega^2 \varphi_2.
\end{aligned}$$

If we now closely examine these functions, we perceive that all values φ_i are obtained from φ_1 by interchanging in all possible ways the letters x_i. Every such permutation changes the function φ_1, and corresponding to the $3! = 6$ possible permutations we have six values φ_i. Hence φ_1 is said to be a six-valued function of the x_i.

The radical $\sqrt{R}$ as function of the x_i is by §6 a multiple of

$$(x_1 - x_2)(x_1 - x_3)(x_2 - x_3).$$

Interchanging here the letters x_i in all possible ways, we obtain just two alternating values with the plus and minus sign, hence $\sqrt{R}$ is said to be a two-valued or **alternating function**[1] of the x_i.

The coefficients of the general cubic equation are as functions of the x_i:

$$\begin{gathered}
x_1 + x_2 + x_3 \\
x_1 x_2 + x_1 x_3 + x_2 x_3 \\
x_1 x_2 x_3.
\end{gathered}$$

Interchanging the letters x_i in all possible ways, we do not alter the value of the coefficients, hence they are said to be one-valued or **symmetric functions**[2] of the x_i.

The function A_i, composed of a one-valued and a two-valued function, is obviously two-valued.

Thus we notice the existence of functions in the roots x_i of the cubic equation which are one-valued, two-valued and six-

[1] Cf. §29.
[2] Cf. §22.

valued with regard to the permutations on these roots. We notice the possibility of computing two-valued functions from one-valued functions by adjoining square roots, and of six-valued functions from two-valued functions by adjoining cube roots. We notice the possibility of expressing the roots x_i themselves rationally in terms of such functions with the highest number of values.

Does all this not suggest a connection between the solution of an equation and the permutations on its roots? It did suggest that to Lagrange.

CHAPTER III

ALGEBRA OF PERMUTATIONS

§9. NOTATION

To interchange symbols is to perform on them a **permutation or substitution.** It is commonly denoted by s or a letter following s.

Cauchy calls such an operation a substitution and the result of it a permutation. But the use of these terms is not settled as yet, and it seems preferable to denote by substitution a different operation, as it occurs in §63.

We shall operate permutations on letters x_i representing roots of an equation; and, more seldom, on functions of these x_i denoting such a permutation by σ.

The n letters x_i of the general case are roots of the general equation

$$f(x) = a_0x^n + a_1x^{n-1} + \ . \ . \ . + a_n = 0.$$

Given the three letters

$$x_1, x_2, x_3,$$

we can perform on them altogether 3! permutations, including the permutation that leaves the arrangement of the letters unaltered. This last one we call the **identical permutation or identity,** and denote by 1 if no confusion can arise.

To agree upon a notation for permutations other than identity, we suppose that some permutation s changes the arrangement

$$x_1, x_2, x_3$$

into the arrangement

$$x_2, x_3, x_1.$$

We could denote this permutation by

$$s = \begin{pmatrix} x_1x_2x_3 \\ x_2x_3x_1 \end{pmatrix},$$

meaning that each letter in the upper row is replaced by the corresponding letter in the lower row. Or, simpler, could denote it by

$$s = \binom{123}{231},$$

writing only the subscripts of the x_i.

But we notice that the permutation s replaces x_1 by x_2, x_2 by x_3, x_3 by x_1; that is, replaces the letters in a cycle. And we readily find a more convenient and frequent notation:

$$s = (x_1x_2x_3),$$

or

$$s = (123),$$

meaning that each letter in the **cycle** is replaced by the following one and the last letter by the first one.

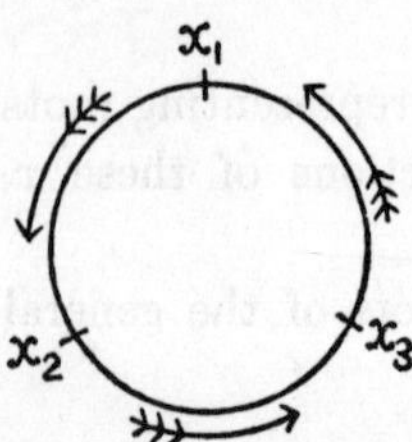

Such a permutation, interchanging the letters x_i cyclically, is called a **circular permutation.** It can suitably be illustrated by a circular diagram, and it should be clear that in the **circular notation** the symbols

$$(123), (231), (312)$$

represent the same permutation, which can be denoted in as many ways as there are letters marked in its cycle.

But not a circular permutation only, any permutation can be represented in the circular notation if we decompose it into cycles. We do that taking any one letter and noting in a cycle the succession of replacing letters until we return to the letter that we started from. Here we close the cycle and begin a new one with a letter not yet noted, and we keep on doing so until the notation is complete. Thus we may have a permutation t on five letters x_i denoted as

$$t = \binom{x_1x_2x_3x_4x_5}{x_3x_4x_5x_2x_1} = (x_1x_3x_5)(x_2x_4) = (135)(24).$$

§10. DEGREE

The number of letters operated on by a permutation is its **degree.** The permutation

$$s = (123)$$

is of degree three, the permutation

$$t = (135)(24)$$

is of degree five. The **degree of a cycle** is the number of letters marked in a cycle. Thus we have to distinguish between the degree of a permutation and the degrees of its cycles: the permutation t, while of degree five, has cycles of degree three and two.

A cycle of degree one notes that the letter marked in the cycle has not been displaced. It is commonly omitted in the notation and not counted toward the degree: we set for instance

$$(135)(2)(4) = (135).$$

A permutation of degree two, as

$$(12),$$

is called a **transposition.** A permutation whose cycles are all of the same degree, as

$$(123)(456),$$

is called **regular.** Two permutations with cycles corresponding as to number and degree, as

$$(123)(45) \text{ and } (135)(24),$$

are called **similar.**

§11. COMBINATION

Permutations obey the **law of combination**: when we have two permutations s and t and apply them successively, we combine or **multiply** them, as we say, into a permutation denoted by st and called the **product** of s and t, and there is only one such product.

For example, to the set of letters

$$x_1, x_2, x_3$$

we apply successively the permutations

$$s = (123)$$

and

$$t = (23).$$

The given arrangement of the x_i is changed by s into

$$x_2, x_3, x_1,$$

and this by t into

$$x_3, x_2, x_1.$$

This result we obtain directly by applying to the given arrangement of the x_i the permutation

$$\begin{pmatrix}123\\321\end{pmatrix} = (13);$$

hence

$$st = (123)(23) = (13).$$

This product of s and t can be read off from the permutations s and t without performing the indicated operations: s replaces 1 by 2 and t replaces 2 by 3, therefore st replaces 1 by 3; s replaces 3 by 1 and t does not alter 1, therefore st replaces 3 by 1 and 3 closes the cycle; s replaces 2 by 3 and t replaces 3 by 2, hence 2 is not displaced by st.

Similarly, in four letters x_i we find that

$$(12)(13)(14) = (1234).$$

Applying first the permutation t and then the permutation s, we obtain the product ts of those permutations which is not the same as the product st:

$$ts = (23)(123) = (12).$$

Hence it appears that permutations do not in general obey the **commutative law.** Yet they may do so, in which case they are called **commutative or permutable**: for instance, the permutations

$$s = (12)(34)$$

and

$$t = (13)(24)$$

on four letters x_i give

$$st = ts = (14)(23).$$

It is obvious that permutations are always commutative when they operate on different letters: for instance, the permutations

$$s = (123)$$

and

$$t = (45)$$

on five letters x_i give

$$st = ts = (123)(45) = (45)(123).$$

It follows that

(13) **a non-circular permutation can always be represented as product of commutative circular permutations.**

§12. ORDER

Multiplying the permutation s by itself, we obtain the products ss, sss, . . . , and adopting a usage of elementary algebra, we call them **powers** of s and set

$$ss = s^2, \quad s^2s = s^3, \ldots$$

If

$$s = (123),$$

then s^2 replaces 1 by 2 and 2 by 3, 3 by 1 and 1 by 2, 2 by 3 and 3 by 1, hence

$$s^2 = (132).$$

Likewise, $s^3 = s^2s$ replaces 1 by 3 and 3 by 1, 2 by 1 and 1 by 2, 3 by 2 and 2 by 3; hence

$$s^3 = (1)(2)(3) = 1.$$

Continuing the involution on s, we make up the table

$$\begin{aligned} 1 &= s^3 = s^6 = \ldots \\ s &= s^4 = s^7 = \ldots \\ s^2 &= s^5 = s^8 = \ldots \end{aligned}$$

and find that we can obtain no other distinct permutations than 1, s, s^2 and that the permutations s^3, s^6, . . . are equal to identity.

The involution on

$$t = (1234)$$

gives

$$\begin{aligned} t^2 &= (13)(24) \\ t^3 &= (1432) \\ t^4 &= (1)(2)(3)(4) = 1. \end{aligned}$$

We notice that the powers of a circular permutation are not all circular unless the degree of the permutation is prime, but they are all regular.

The lowest power of a permutation or a cycle that equals identity, or the difference of two successive powers that are equal, is the **order or period** of the permutation or the cycle.

Thus the permutation $s = (123)$ is of order three, the permutation $t = (1234)$ is of order four.

To form the square of a circular permutation we replace each letter by the second to its right, to form the cube, by the third. If we have a circular permutation s of degree n, the n-th power of s, replacing each letter by the n-th to its right, replaces each letter by itself and is identity.

Any power of such a permutation s higher than n equals some power of s lower than n, and all distinct permutations that we can obtain by involution on s are contained in the set

$$1, s, s^2, \ldots, s^{n-1},$$

or

$$s, s^2, s^3, \ldots, s^n = 1.$$

Hence it appears that

(14) **the order of a circular permutation is equal to its degree, and both are determined by the number of letters acted upon by the permutation.**

To find the order of a non-circular permutation, we decompose it into cycles; to the cycles we then apply the proposition for circular permutations. For instance, the permutation

$$u = (123)(45)$$

gives

$$u^2 = (132)(4)(5), \quad u^3 = (1)(2)(3)(45),$$
$$u^4 = (123)(4)(5), \quad u^5 = (132)(45),$$
$$u^6 = (1)(2)(3)(4)(5) = 1$$

and is of order six.

Clearly then,

(15) **the order of a non-circular permutation is the lowest common multiple of the orders of its cycles,**

for only such a power of a non-circular permutation resolved into cycles turns each cycle into identity.

It follows that the order of a permutation cannot be prime unless the permutation is circular of prime degree or regular with cycles of prime degree. Such permutations are, for instance:

$$s = (123) \qquad v = (123)(456)$$
$$s^2 = (132) \qquad v^2 = (132)(465)$$
$$s^3 = 1 \qquad v^3 = 1.$$

Hence we can note:

(16) **If the order of a permutation is prime, its powers other than identity are similar permutations,**

which is readily verified. Furthermore:

(17) **If order and degree of a permutation are prime, the permutation is circular,**

for a regular permutation with several cycles cannot be of prime degree.

§13. ASSOCIATION

While the commutative law does not generally hold for permutations, the **associative law** does:

$$(st)u = s(tu),$$

and we write for both stu. Likewise, we have

$$(st)(uv) = (stu)v = s(tuv) = s(tu)v = stuv.$$

We can test the associative law on any example, but it also follows for permutations from the law of combination. This law, requiring that the product of two permutations be identically defined, can be extended to the product of three permutations. But the simplest way of expressing the postulate that such a product be identically defined is the associative law.

From the associative law follows that

$$s^k s^l = s^l s^k = s^{k+l},$$

as we see from the scheme

$$\rlap{\overbrace{}^{k}\;\overbrace{}^{l}}\underbrace{ss\ldots s}_{l}\;\underbrace{s\ldots s\;s\ldots s.}_{k}$$

Hence powers of a permutation are commutative.

If the two permutations s and t are commutative, we can set

$$s^k t^l = t^l s^k.$$

This we show to be true for the particular case s^3t^2, by induction it will then be true in general:

$$s^3t^2 = ssstt = sstst = ststs = tstss = ttsss = t^2s^3.$$

We can also set

$$(st)^2 = s^2t^2$$

if the permutations s and t are commutative; but not otherwise, since

$$(st)^2 = stst = sstt = s^2t^2.$$

To set

$$(stu\ .\ .\ .\ )^2 = s^2t^2u^2\ .\ .\ .\ ,$$

it is necessary that every pair of permutations be commutative.

§14. INVERSE

There always exists a permutation that undoes or reverses the interchange of letters effected by another permutation, and it is called the **inverse** of that permutation. The product of a permutation and its inverse evidently is identity. Again, if the product of two permutations is identity, each permutation is inverse to the other. For instance, the permutations

$$s = \begin{pmatrix} 123 \\ 231 \end{pmatrix} = (123)$$

and

$$t = \begin{pmatrix} 231 \\ 123 \end{pmatrix} = (132)$$

are each inverse to the other, and

$$st = 1.$$

It is clear that we obtain the inverse of a permutation by reversing the order of the numbers in its cycles. The example makes this evident for circular permutations:

$$s = (123)$$
$$t = (321),$$

and for non-circular permutations it follows by proposition (13). We infer that

(18) **the inverse of a permutation s is a permutation similar to s.**

If the order of a permutation s is r, the permutations s^k and s^{r-k} are each inverse to the other because

$$s^k s^{r-k} = 1.$$

If the order is three, s^2 and s are each inverse to the other; if the order is two, s is its own inverse.

Following a usage of elementary algebra, we denote the inverse of a permutation s by s^{-1}, so that

$$ss^{-1} = 1$$

and

$$t = s^{-1}$$

if

$$st = 1.$$

We adopt also the elementary notation

$$s^{-1}s^{-1} = s^{-2}, \quad s^{-2}s^{-1} = s^{-3}, \ldots$$

and the symbols s^1 and s^0 defined by the relations

$$s^1 = s$$

and

$$s^0 = 1,$$

so that

$$ss^{-1} = s^1s^{-1} = s^0 = 1.$$

The inverse of s^2 is s^{-2} because

$$s^2s^{-2} = s(ss^{-1})s^{-1} = 1,$$

and a like relation is true for any power.

The inverse of st, denoted by $(st)^{-1}$, is $t^{-1}s^{-1}$ and not $s^{-1}t^{-1}$, for only

$$st \, . \, t^{-1}s^{-1} = 1;$$

having operated the permutation st we must retrace our way back step by step.

With the indices, we define also **division** by a permutation as elementary algebra would have it, namely as multiplication by the inverse of the permutation. Thus

$$\frac{s^3}{s^2} = s^3 \, . \, s^{-2} = s.$$

If

$$s^3 = 1,$$

multiplication by s is equivalent to division by s^2; which we readily interpret on the circular diagram of §9, taking division and the negative sign of an exponent as reversing direction.

§15. IDENTITY

The multiplication of a permutation on either side by identity obviously does not give a product different from that permutation:

$$1 \,.\, s = s \,.\, 1 = s.$$

Identity may therefore be regarded as commutative with every permutation and may be suppressed in the notation of a product.

Conversely,

(19) if *e* and *s* are two permutations such that

$$\boxed{es = se = s},$$

then *e* is identity.

For

$$1 \,.\, e = e$$

from the nature of identity, and

$$1 \,.\, e = 1$$

since it is given that $se = s$. Hence it follows that

$$e = 1.$$

We conclude with a few examples:

(a) Prove that $t = u$ if $st = su$:

$$s^{-1} \,.\, (st) = s^{-1} \,.\, (su); \quad (s^{-1}s)t = (s^{-1}s)u; \quad t = u.$$

This verifies that $e = 1$ if $es = se = s$, for $s1 = s$ and $se = s$ give $se = s1$ and consequently $e = 1$.

(b) Prove that st is the inverse of ts if $s^2 = t^2 = 1$:

$$st \,.\, ts = st^2s = s^2 = 1$$

It follows that

$$st = (ts)^{-1} = s^{-1}t^{-1}.$$

(c) Prove that s and t are commutative if $s^2 = t^2 = (st)^2 = 1$:

$$st = t^2st = ttst = ts^2tst = tsstst = ts(st)^2 = ts.$$

(d) Prove that $h_2 = s^{-1}h_1s$ if $sh_2s^{-1} = h_1$:

$$sh_2s^{-1} = h_1; \quad s^{-1}sh_2s^{-1} = s^{-1}h_1; \quad h_2s^{-1}s = s^{-1}h_1s; \quad h_2 = s^{-1}h_1s.$$

CHAPTER IV

GROUP AND SUBGROUP

§16. GROUP

Suppose we have n letters x_i. By functions of such letters we always mean rational functions, both here and later.

If to a function φ of the x_i we apply all permutations that are possible between the x_i, there will be some that do not alter the function φ. Let them compose the set

$$s_1, s_2, \ldots, s_r.$$

A complete set of distinct permutations that do not alter a function is called a **group of permutations** and is denoted by G unless we use special notation. It will often be found convenient to mark G as a group by noting $\{G\}$. The development of the group theory is due to the genius of Cauchy.[1]

A function φ is said to belong to a group of permutations if it remains unaltered by all those, and only those, permutations which are in the group. It appears that

(20) **every function φ of the x_i belongs to a group G of permutations between the x_i.**

Since identity is contained in every group of permutations, we commonly set

$$G = 1, s_2, \ldots, s_r,$$

assuming that

$$s_1 = 1.$$

When we apply two permutations of G successively to the function φ, or one permutation twice, the operation leaves φ unaltered since every permutation of G does so.

It follows that a group of permutations contains the product of any permutation by itself or another permutation, hence in

[1] Cauchy lived 1789–1857.

particular the inverse of any permutation. And this property defines a group if we can prove that

(21) **to every group G of permutations between the x_i belongs a function φ of the x_i.**

The function

$$\psi_1 = \alpha_1 x_1 + \alpha_2 x_2 + \ldots + \alpha_n x_n$$

is evidently altered by every permutation of G other than identity, and by no two permutations alike. Suppose that the permutations

$$1, s_2, \ldots, s_r$$

of G convert ψ_1 into

$$\psi_1, \psi_2, \ldots, \psi_r$$

respectively. We form the function

$$\varphi = \psi_1 + \psi_2 + \ldots + \psi_r;$$

this function belongs to G and satisfies the proposition.

For any permutation s_i of G applied to φ gives

$$\varphi_i = \psi_{1i} + \psi_{2i} + \ldots + \psi_{ri},$$

where the subscripts identify the permutations that have been applied to the functions. But each of the permutations

$$1s_i, s_2 s_i, \ldots, s_r s_i$$

is in G since it is the product of two permutations which are in G, and no two of these permutations are alike since from

$$s_j s_i = s_k s_i$$

would follow

$$s_j = s_k,$$

which is untrue. These permutations represent therefore in some order or other the permutations of G, and hence the ψ_{ki} in some order or other the ψ_i. Consequently

$$\varphi_i = \varphi.$$

If a permutation t is not in G, then $s_i\, t$ is not in G either, and

$$\varphi_t \neq \varphi.$$

The function φ, remaining unaltered by every permutation of G and such a permutation alone, belongs to G. And since we can construct in the way indicated any number of functions belonging to G,

(22) **a group of permutations is defined as a set of distinct permutations such that the product of any two, or the square of any one, is a permutation of the set.**

The number of permutations contained in a group is its **order,** the number of letters that its permutations operate on is its **degree.**

The degree of a group G is denoted by n which may be added to the symbol of the group as in G^n.

The order of a group G is denoted by r, or by r_g to identify the group it belongs to, and may be added to the symbol of the group as in G_r.

The simplest group in a sense is **identity** by itself. It is denoted by 1 if no confusion can result, otherwise by $\{1\}$, and its order is one. In ψ_1 we had a function belonging to identity.

Next in simplicity is a group containing only the powers of a permutation s. It is called a **cyclic group** and denoted by C or

$$\{s\} = s, s^2, \ldots s^r;$$

its order is the order of the permutation s.

The complete group of all $n!$ permutations that are possible between n letters x_i is called the **symmetric group** on these letters. It is the group of symmetric functions,[1] is denoted by S, and $n!$ is its order.

§17. SUBGROUP

A group which is contained in another group is called its **subgroup**, and we denote a group and its subgroup by G and H respectively when we do not use special notation.

Such groups stand in a beautiful relation. Let the permutations of the set

$$H = 1, s_2, \ldots, s_r$$

form a subgroup of order r in G and let ψ be a function that belongs to H. If t is a permutation contained in G but not in H and converts ψ into ψ_t such that

$$\psi_t \neq \psi,$$

[1] Cf. §22; also §8.

then every permutation of the set

$$Ht = t, s_2t, \ldots, s_rt$$

converts ψ into ψ_t, for s_i leaves ψ unaltered and t effects the change.

There is no permutation outside Ht that converts ψ into ψ_t. Suppose that τ does so; then τt^{-1} leaves ψ unaltered as t^{-1} converts ψ_t into ψ. Hence

$$\tau t^{-1} = s_i,$$

and after multiplication by t on the right:

$$\tau = s_i\, t.$$

The permutations of Ht are all different from the permutations of H, and they are all unlike since from

$$s_i\, t = s_k\, t$$

would follow

$$s_i = s_k.$$

If the sets of H and Ht do not contain all the permutations of G, there is a permutation u in G but neither in H nor in Ht that converts ψ into ψ_u. And so does every permutation of the set

$$Hu = u, s_2u, \ldots, s_ru,$$

to which all conclusions for Ht apply.

Continuing so until the group G is exhausted:

$$\begin{aligned} H &= 1, s_2, \ldots, s_r && [\psi \rightarrow \psi \\ Ht &= t, s_2t, \ldots, s_rt && [\psi \rightarrow \psi_t \\ Hu &= u, s_2u, \ldots, s_ru && [\psi \rightarrow \psi_u \\ &\ldots\ldots\ldots, \end{aligned}$$

we arrange the permutations of

$$G = H + Ht + Hu + \ldots$$

in **partitions** of G with respect to H or **co-sets** of H in G, as we say; and they convert ψ into

$$\psi, \psi_t, \psi_u, \ldots$$

respectively, which functions are called **conjugate** with ψ under G.

We may arrange the permutations so as to set

$$G = H + tH + uH + \ldots,$$

but in general

$$Ht \neq tH,$$

for the commutative law does not always hold for permutations and sets of such.[1]

Since every partition contains as many permutations as H does, and all partitions together contain as many permutations as G does, the proposition follows which is due to Lagrange:

(23) **The order of a group is divisible by the order of any subgroup.**

If r_g denotes the order of G and r_h the order of H, then

$$r_g = j \cdot r_h,$$

where j is called the **index** of H in G. It gives the number of partitions in G with respect to H and the number of conjugate values that a function ψ belonging to H takes under G.

Any group on the n letters x_i is contained in the symmetric group S on these letters. Denoting the index of G in S by j_g and the index of H in S by j_h, we have therefore

$$n! = j_g r_g = j_h r_h,$$

so that the index j or, better, j_{hg} of H in G is expressible as a quotient of either orders or indices:

$$j_{hg} = \frac{r_g}{r_h} = \frac{j_h}{j_g}.$$

Since the order of any possible subgroup H of S has to divide $n!$, there are for $n = 4$, let us say, no other subgroups of S possible than such as have

$$\begin{array}{lcccccccc} r_h = & 1 & 2 & 3 & 4 & 6 & 8 & 12 & 24 \\ j_h = & 24 & 12 & 8 & 6 & 4 & 3 & 2 & 1. \end{array}$$

As illustration may serve an example:

$H = 1, (12), (34), (12)(34), (13)(24), (14)(23), (1324), (1423)$

$$\psi = x_1x_2 + x_3x_4$$

$Ht = (23), (132), (234), (1342), (1243), (14), (124), (143)$

$$\psi_t = x_1x_3 + x_2x_4 \qquad [t = (23)$$

$Hu = (24), (142), (243), (1432), (13), (1234), (134), (123)$

$$\psi_u = x_1x_4 + x_2x_3. \qquad [u = (24)$$

[1] By §11. But cf. proposition (32): $Nt = tN$; cf. also proposition (82).

In a sense every group G contains itself and identity as subgroups, but it often is more convenient to consider one or both as not included among them. If this is not specified, it will be clear from the context.

Since the powers of a permutation s in G by themselves compose a cyclic group $\{s\}$ and this group is a subgroup of G,

(24) **the order of any permutation in a group divides the order of the group.**

If the order of a group G is prime, it follows that the powers of a permutation s in G compose the whole group which then is cyclic:

$$G = \{s\};$$

and since s is any permutation in G, that

$$\{s\} = \{s^k\},$$

which becomes intelligible if we recall that by proposition (16) the powers of s are similar permutations: the order of s^k equals the order of s while every power of s^k equals some power of s.

Calling a cyclic group $\{s\}$ **circular** when s is so, we note that

(25) **a group of prime order is cyclic, a group of prime order and degree is circular;**

the latter by proposition (17).

§18. CONJUGATE SUBGROUPS

While the permutations

$$Ht = t, s_2 t, \ldots, s_r t$$

of G convert the function ψ belonging to

$$H = 1, s_2, \ldots, s_r$$

into the conjugate function ψ_t and compose a partition of G, they do not compose the group of ψ_t; not even a group at all since we notice that they do not contain identity and contain t but none of its powers.

To find the group of ψ_t which we denote by H_t, we let τ be any permutation contained in it, so that τ applied to ψ_t leaves it unaltered:

$$\psi_{t\tau} = \psi_t.$$

To both sides of this identity we apply the permutation t^{-1} and obtain

$$\psi_{t\tau t^{-1}} = \psi_{tt^{-1}} = \psi$$

since the permutation $tt^{-1} = 1$ does not alter ψ. It follows that the permutation $t\tau t^{-1}$ does not alter ψ either and is some permutation of H:

$$t\tau t^{-1} = s_i.$$

Hence we find

$$\tau = t^{-1}s_i\, t$$

as shown in example (d) of §15.

The permutation $t^{-1}s_i\, t$ is called the **transform of permutation** s_i by t. Any such transform leaves ψ_t unaltered as t^{-1} changes ψ_t into ψ, and s_i leaves ψ fixed, and t changes ψ back into ψ_t. Since any permutation τ of H_t is the transform of a permutation s_i in H by t, and any transform of a permutation s_i in H by t leaves ψ_t unaltered, we infer that the group of ψ_t is

$$H_t = 1, t^{-1}s_2\, t, \ . \ . \ . \ , t^{-1}s_r\, t. \qquad [t^{-1}s_1\, t = 1$$

This set of transforms of the permutations in H by t is called the **transform of group** H by t and denoted[1] by $t^{-1}Ht$; hence we can set

$$H_t = t^{-1}Ht.$$

It is clear that H_t is a subgroup of G. Since the permutations of H are all distinct, their transforms are so, too, and H_t must have the same order and index in G that H has.

As the function ψ_t is called conjugate with ψ under G, so the permutation τ is called **conjugate** with s_i under G, and the subgroup H_t is called **conjugate** with H under G.

It follows that

(26) **the transform of a permutation is a conjugate permutation, the transform of a subgroup is a conjugate subgroup,**

and that

(27) **conjugate subgroups have the same order and index.**

Corresponding to the partitions

$$H, Ht, Hu, \ . \ . \ .$$

[1] Sometimes by tHt^{-1}, which is read from the right.

and the conjugate functions

$$\psi, \psi_t, \psi_u, \ldots$$

we thus have the conjugate subgroups

$$H, H_t, H_u, \ldots$$

such that

$$H_t = t^{-1}Ht$$
$$H_u = u^{-1}Hu$$
$$\cdot \quad \cdot \quad \cdot \quad \cdot \quad \cdot \quad \cdot \quad \cdot$$

This gives the proposition:

(28) **If j is the index of H in G, a function ψ belonging to H takes j conjugate values under G which belong to subgroups conjugate with H under G.**

Since the product of any two permutations s_i and s_j of H is equal to a permutation s_k contained in H, the product of their transforms by any permutation t of G is

$$t^{-1}s_i\, t \,.\, t^{-1}s_j\, t = t^{-1}s_i\, s_j\, t = t^{-1}s_k\, t$$

contained in $t^{-1}Ht$. This verifies that the transform of H by t is a group.

The transform of a product equals the product of the transforms, for

$$t^{-1}s_i\, s_j\, t = t^{-1}s_i\, t \,.\, t^{-1}s_j\, t,$$

whence the transform of a power equals that power of the transform:

$$t^{-1}s_i^2\, t = (t^{-1}s_i\, t)^2.$$

The transform of the inverse equals the inverse of the transform:

$$t^{-1}s_i^{-1}\, t = (t^{-1}s_i\, t)^{-1},$$

for

$$t^{-1}s_i^{-1}\, t \,.\, t^{-1}s_i\, t = 1.$$

§19. RULE OF TRANSFORMS

To avoid the multiplication of three permutations in computing transforms, we make use of a simple device.

Suppose we have the permutations

$$s = (abcd)$$

and

$$t = \begin{pmatrix} abcd\ldots \\ klmn\ldots \end{pmatrix},$$

the letters in parenthesis denoting the subscripts of the x_i upon which the permutations s and t operate. The inverse of t is

$$t^{-1} = \begin{pmatrix} klmn \ldots \\ abcd \ldots \end{pmatrix},$$

and the transform $t^{-1}s\,t$ replaces k by a by b by l, l by b by c by m, m by c by d by n, n by d by a by k, which closes the cycle:

$$t^{-1}s\,t = (klmn).$$

But this result we obtain also if we replace the subscripts in the cycle of the permutation s as indicated by the permutation t.

When s is not a circular permutation, we represent it by proposition (13) as product of circular permutations s_1 and s_2, say, decomposing it into cycles. We then perform the operation on every cycle, which we can do since

$$t^{-1}s\,t = t^{-1}s_1s_2\,t = t^{-1}s_1\,t \,.\, t^{-1}s_2\,t.$$

Hence we have the **rule of transforms:**

(29) **The transform of a permutation s by a permutation t is obtained by operating t within the cycles of s;**

and it is clear that

(30) **the transform of a permutation s is a permutation similar to s.**

The order of any transform of a permutation is therefore the same as that of the permutation itself.

An example may illustrate the rule:

H = 1, (12), (34), (12)(34), (13)(24), (14)(23), (1324), (1423)

$$\psi = x_1x_2 + x_3x_4$$

H_t = 1, (13), (24), (13)(24), (12)(34), (14)(32), (1234), (1432)

$$\psi_t = x_1x_3 + x_2x_4 \qquad [t = (23)$$

H_u = 1, (14), (32), (14)(32), (13)(42), (12)(43), (1342), (1243)

$$\psi_u = x_1x_4 + x_3x_2. \qquad [u = (24)$$

This example makes the rule of transforms self-evident: if we convert ψ into ψ_t by interchanging x_2 and x_3, we have only to interchange x_2 and x_3 in the permutations of the group to which ψ belongs in order to obtain the group of ψ_t.

It will be noticed that the permutations of conjugate subgroups do not make up a group:

$$G = H + Ht + Hu + \ldots$$

but

$$G \neq H + H_t + H_u + \ldots$$

This may be verified by taking as G the symmetric group on four letters given in §21 and as H the group above.

§20. NORMAL SUBGROUP

If a subgroup of G is identical with the subgroups conjugate to it under G, so that transforming it by any permutation of G we reproduce the same subgroup, the subgroup is called **normal or invariant or self-conjugate**[1] in G and is denoted by N or J.

If N is a normal subgroup of G, we have therefore

$$t^{-1}Nt = N$$

for any permutation t of G, whence N contains with a permutation s also every transform of s by a permutation of G. These transforms are by proposition (30) permutations similar to s; yet N may contain not all permutations of G similar to s. For instance, the group

$$G = 1, (12), (34), (12)(34)$$

has a normal subgroup

$$N = 1, (12)$$

not containing the permutation (34) similar to (12).

But every permutation similar to s is its transform by some other permutation, and that permutation is by necessity contained in the symmetric group; consequently

(31) **a normal subgroup of the symmetric group contains with a permutation *s* also every permutation of that group which is similar to *s*,**

therefore every possible such permutation.

From the relation

$$t^{-1}Nt = N$$

we obtain the other:[2]

$$Nt = tN,$$

[1] Normal subgroups were discovered by Galois.

[2] Cf. §17, where $Ht \neq tH$. Cf. also proposition (82).

and conversely. This means that

(32) **every permutation of a group is commutative with a subgroup when, and only when, the subgroup is normal.**

A subgroup which is normal in the group G and contained in the subgroup H of G is evidently normal also in H, but a subgroup which is normal in H is not necessarily normal in G. Thus the group N of the example above is not normal in the symmetric group.[1]

Every group G may be taken to contain itself and identity as normal subgroups, because for any permutation t of G we have

$$t^{-1}1t = 1$$

and

$$t^{-1}Gt = G.$$

But we do not ordinarily mean these groups when we speak of normal subgroups.

A group containing no normal subgroup except itself and identity is called **simple**; a group which is not simple is **composite.**

A group of prime order always is simple, since it can have no subgroups whatever different from itself and identity.

Normal subgroups of their symmetric groups are, for instance:

(a) $N = 1, (123), (132)$

$$\psi = (x_1 + \omega x_2 + \omega^2 x_3)^3$$

$N_t = 1, (132), (123) = N$

$$\psi_t = (x_1 + \omega^2 x_2 + \omega x_3)^3. \qquad [t = (23)$$

(b) $N = 1, (12)(34), (13)(24), (14)(23)$

$$\psi = (x_1 - x_2)(x_3 - x_4)$$

$N_t = 1, (13)(42), (14)(32), (12)(34) = N$

$$\psi_t = (x_1 - x_3)(x_4 - x_2) \qquad [t = (234)$$

$N_u = 1, (14)(23), (12)(43), (13)(42) = N$

$$\psi_u = (x_1 - x_4)(x_2 - x_3). \qquad [u = (243)$$

[1] Another example is in §46.

CHAPTER V

SYMMETRIC GROUP AND ITS FUNCTIONS

§21. GENERATOR

The permutations of a set are called **independent** if no one of them can be expressed as a product of the rest. They define a group and as such they are called **generating permutations.** To obtain that group, we combine the generators in all possible ways until we get only such permutations as we already have.

The $n - 1$ independent transpositions

$$(x_1x_i)_2^n = (x_1x_2), (x_1x_3), \ldots, (x_1x_n)$$

generate the symmetric group on the n letters x_i. To verify this, we consider that every permutation is composed of cycles or resolvable into such by proposition (13); that every cycle is decomposable into transpositions:

$$(x_1x_2x_3x_4) = (x_1x_2)(x_1x_3)(x_1x_4);$$

and that every transposition is expressible as product of some transpositions in $(x_1x_n)_2^n$:

$$(x_2x_3) = (x_1x_2)(x_1x_3)(x_1x_2).$$

Since there is no permutation in the symmetric group which is not a combination of transpositions in $(x_1x_i)_2^n$, we infer that all possible combinations of these transpositions give us the symmetric group.

If $\{s,t\}$ stands for the group generated by the permutations s and t, we can note therefore that

(33) **the symmetric group on n letters x_i is generated by the $n - 1$ independent transpositions $(x_1x_i)_2^n$:**

$$\boxed{S^n = \{(x_1x_i)_2^n\}}.$$

It follows that a group is symmetric whenever it contains all transpositions $(x_1x_i)_2^n$.

For the lowest degrees the symmetric groups are:

$S^2 = 1, (12)$

$S^3 =$

1	(13)	(23)
(12)	(123)	(132)

$S^4 =$

1	(14)	(24)	(34)
(12) (13) (23) (123) (132)	(124) (134) (14)(23) (1234) (1324)	(142) (13)(24) (234) (1423) (1342)	(12)(34) (143) (243) (1243) (1432)

As to the rule by which to form a symmetric group of higher degree from the preceding symmetric group of lower degree, we notice that

$$S^3 = S^2 + S^2t_1 + S^2t_2,$$

where

$$t_1 = (13) \text{ or } (123)$$
$$t_2 = (23) \text{ or } (132);$$

and

$$S^4 = S^3 + S^3t_1 + S^3t_2 + S^3t_3,$$

where

$$t_1 = (14) \text{ or } (1234)$$
$$t_2 = (24) \text{ or } (13)(24)$$
$$t_3 = (34) \text{ or } (1432).$$

In general

$$S^n = S^{n-1} + S^{n-1}t_1 + S^{n-1}t_2 + \cdots + S^{n-1}t_{n-1},$$

where

$$t_1 = (1n) \text{ or } (1 \ldots n)$$
$$t_2 = (2n) \text{ or } (1 \ldots n)^2$$
$$\cdots\cdots\cdots$$
$$t_{n-1} = ([n-1]n) \text{ or } (1 \ldots n)^{n-1}.$$

§22. SYMMETRIC SUM

Any function of the n letters x_i that remains unaltered by the permutations of the symmetric group on these letters is called a **symmetric function** of the x_i and denoted by S. It remains unaltered because the permutations of the symmetric group do nothing but interchange its terms.[1]

A symmetric function does not have to be homogeneous in the x_i. But if it is not, we evidently can express it as a sum of homogeneous functions collecting the terms of one total degree[2] in the x_i into one such function.

Every homogeneous function which is part of a symmetric function is symmetric itself because permutations, interchanging the terms of the symmetric function, interchange the terms within every homogeneous function only, since they cannot alter the total degree of a term.

If a homogeneous symmetric function contains the term

$$a \,.\, x_1^{\mu_1} x_2^{\mu_2} \ldots x_k^{\mu_k},$$

it also contains every term that is obtained applying to this the permutations of the symmetric group which interchange the subscripts of the x_i leaving the exponents fixed.

The sum of terms obtained in such a manner we call a **symmetric sum** and denote by s. With the term that we noted the homogeneous symmetric function contains the symmetric sum

$$s(a \,.\, x_1^{\mu_1} \ldots x_k^{\mu_k}) = a \,.\, s(x_1^{\mu_1} \ldots x_k^{\mu_k});$$

the notation will be readily understood. This symmetric sum has $k!$ terms if all the exponents μ_i are distinct; if some of them are alike, including those that may be zero, the number of terms is a fraction only of $k!$.

A term of the homogeneous symmetric function which is not contained in this symmetric sum belongs to another, so that a homogeneous symmetric function is composed of symmetric sums. They are all of the same total degree in any one term since they belong to one homogeneous function.

[1] This applies to integral symmetric functions; fractional symmetric functions are expressible as quotients of integral symmetric functions, which alone we consider.

[2] Cf. §24.

§23. COMPUTATION OF SYMMETRIC SUM

We prove that

(34) **a rational symmetric sum in the n letters x_i which represent the roots of the general equation**

$$\boxed{f(x) = x^n - c_1x^{n-1} + \ldots \pm c_n = 0}$$

is rationally expressible in terms of the elementary symmetric functions

$$\boxed{\begin{aligned} s(x_1) &= x_1 + x_2 + x_3 + \ldots = c_1 \\ s(x_1x_2) &= x_1x_2 + x_1x_3 + \ldots = c_2 \\ &\ldots\ldots\ldots \\ s(x_1 \ldots x_n) &= x_1x_2x_3 \ldots x_n = c_n \end{aligned}}.$$

To obtain a unique arrangement of terms in any function, we agree to call the first of two terms

$$p \,.\, x_1^{\mu_1}x_2^{\mu_2} \ldots x_k^{\mu_k}$$

and

$$q \,.\, x_1^{\nu_1}x_2^{\nu_2} \ldots x_l^{\nu_l}$$

higher if $\mu_1 > \nu_1$ or, μ_1 being equal to ν_1, if $\mu_2 > \nu_2$, and so on. This convention will hold also in the case that some x_i has the exponent zero and disappears.

To obtain a unique arrangement of functions, like symmetric sums, we agree to call the one higher whose highest term is higher.

Two terms could be equally high only if all corresponding exponents were equal, but would for that same reason be added into one term. Likewise, two equally high symmetric sums would be added into one sum.

The highest terms of the elementary symmetric functions

$$s(x_1),\ s(x_1x_2),\ s(x_1x_2x_3),\ \ldots$$

are respectively

$$x_1, \quad x_1x_2, \quad x_1x_2x_3, \ \ldots$$

The highest term of the homogeneous symmetric function

$$c_1^{\mu_1-\mu_2}c_2^{\mu_2-\mu_3} \ldots c_{k-1}^{\mu_{k-1}-\mu_k}c_k^{\mu_k} = C$$

of the x_i is obtained as the product of the highest terms in its factors if we assume that their exponents are not negative:

$$\mu_1 - \mu_2 \geqq 0,\ \mu_2 - \mu_3 \geqq 0,\ \ldots$$

It is

$$x_1^{\mu_1-\mu_2}(x_1x_2)^{\mu_2-\mu_3} \ldots (x_1 \ldots x_k)^{\mu_k} = x_1^{\mu_1}x_2^{\mu_2} \ldots x_k^{\mu_k},$$

so that the highest symmetric sum in C is

$$s(x_1^{\mu_1}x_2^{\mu_2} \ldots x_k^{\mu_k}).$$

It follows that $x_1^{\mu_1} \ldots x_k^{\mu_k}$, being the highest term in C, also is the highest term in $s(x_1^{\mu_1} \ldots x_k^{\mu_k})$, with which we readily agree recalling the assumption as to the exponents and writing it in the form

$$\mu_1 \geqq \mu_2 \geqq \ldots \geqq \mu_k.$$

With this assumption made, we prove our proposition for $s(x_1^{\mu_1}x_2^{\mu_2} \ldots x_k^{\mu_k})$ on the lines of actual computation.

Subtracting s from C we obtain

$$C - s = S,$$

where

$$S = \sum_1^m p_i s_i\,(x_1^{\mu_{i1}}\,x_2^{\mu_{i2}} \ldots x_{k_i}^{\mu_{ik_i}})$$

is a homogeneous symmetric function composed of symmetric sums s_i all lower than s and arranged so that every s_i is higher than s_{i+1} while every

$$\mu_{ij} \geqq \mu_{ij+1}.$$

For the lowest possible sum s_m with

$$\mu_{m1} = \mu_{m2} = \ldots = \mu_{mk_m} = 1$$

and

$$k_m = n$$

we have the total degree

$$\sum \mu_i = \sum \mu_{ij} = n.$$

From S we subtract

$$p_1C_1 = p_1c_1^{\mu_{11}-\mu_{12}} \ldots c_{k_1}^{\mu_{1k_1}}$$

with the highest term $p_1x_1^{\mu_{11}} \ldots x_{k_1}^{\mu_{1k_1}}$ and the highest symmetric sum

$$p_1s_1(x_1^{\mu_{11}} \ldots x_{k_1}^{\mu_{1k_1}}).$$

This removes p_1s_1 from S and gives

$$S - p_1C_1 = S_1,$$

where

$$S_1 = \sum_{2}^{m} q_i s_i \, (x_1^{\mu_{i1}} \ldots x_{k_i}^{\mu_{ik_i}})$$

is composed of symmetric sums all lower than s_1. With S_1 we now proceed as we did with S.

Having eliminated successively the highest remaining symmetric sums, we rewrite our equations of elimination

$$s = C - S$$
$$0 = S - p_1 C_1 - S_1$$
$$\cdot \quad \cdot \quad \cdot \quad \cdot \quad \cdot$$

and adding them obtain

$$s = C - p_1 C_1 \pm q_2 C_2 \pm \ldots,$$

which proves our proposition.

Since any symmetric function is composed of homogeneous symmetric functions, any homogeneous symmetric function is composed of symmetric sums, and any symmetric sum is rationally expressible in terms of the elementary symmetric functions, we have the **fundamental theorem of symmetric functions:**

(35) **Every rational symmetric function is rationally expressible in terms of the elementary symmetric functions.**

While the procedure of the proof may seem complicated, it will become clear from examples:

(a) $s(x_1^2 x_2^2)$.

$$\begin{aligned}
S &= c_2^2 - s(x_1^2x_2^2) = (x_1x_2 + \ldots + x_3x_4)^2 - s(x_1^2x_2^2) \\
&= [s(x_1^2x_2^2) + 2s_1(x_1^2x_2x_3) + 6s_2(x_1x_2x_3x_4)] - s(x_1^2x_2^2) \\
&= 2s_1(x_1^2x_2x_3) + 6s_2(x_1x_2x_3x_4) \\
S_1 &= S - 2c_1c_3 = S - 2[s_1(x_1^2x_2x_3) + 4s_2(x_1x_2x_3x_4)] \\
&= 2s_2(x_1x_2x_3x_4) = 2c_4 \\
s &= c_2^2 - 2c_1c_3 + 2c_4.
\end{aligned}$$

To expand

$$c_2^2 = (x_1x_2 + \ldots + x_3x_4)^2,$$

we observe that the term $x_1^2x_2^2$ can appear only once, as the product $x_1x_2 \, . \, x_1x_2$, this determining the coefficient of s in the expansion. The term $x_1^2x_2x_3$ twice, as product $x_1x_2 \, . \, x_1x_3$ and $x_1x_3 \, . \, x_1x_2$, which gives the coefficient of s_1. The term $x_1x_2x_3x_4$ six times, as product $x_1x_2 . x_3x_4$, $x_1x_3 . x_2x_4$, $x_1x_4 . x_2x_3$, $x_2x_3 . x_1x_4$, $x_2x_4 . x_1x_3$, $x_3x_4 . x_1x_2$, which gives the coefficient of s_2.

To expand

$$c_1c_3 = (x_1 + \ . \ . \ . \ + x_4)(x_1x_2x_3 + \ . \ . \ . \ + x_2x_3x_4),$$

we notice that the term $x_1^2x_2x_3$ appears once, as the product $x_1 \cdot x_1x_2x_3$, this determining the coefficient of s_1; the term $x_1x_2x_3x_4$ four times, as product $x_1 . x_2x_3x_4$, $x_2 . x_1x_3x_4$, $x_3 . x_1x_2x_4$, $x_4 . x_1x_2x_3$, which gives the coefficient of s_2.

(b) $s(x_1^2x_2^2x_3)$.

$$S = c_2c_3 - s = 3s_1(x_1^2x_2x_3x_4) + 10\ s_2(x_1x_2x_3x_4x_5)$$

$$S_1 = S - 3c_1c_4 = 5s_2(x_1x_2x_3x_4x_5) = 5c_5$$

$$s = c_2c_3 - 3c_1c_4 + 5c_5.$$

Since the highest subscript of the c_i determines the degree of the general equation whose coefficients they are and the number of its roots x_i, the highest subscript of the c_i may not be greater than the number of the x_i in the symmetric sum s, and higher subscripts must not appear in the computation. Marking the number of the x_i in s by an index in parenthesis, we have for instance:

(a) $s^{(2)}(x_1^2x_2^2) = c_2^2$; $s^{(3)}(x_1^2x_2^2) = c_2^2 - 2c_1c_3$;

$$s^{(4)}(x_1^2x_2^2) = c_2^2 - 2c_1c_3 + 2c_4.$$

(b) $s^{(3)}(x_1^2x_2^2x_3) = c_2c_3$; $s^{(4)}(x_1^2x_2^2x_3) = c_2c_3 - 3c_1c_4$;

$$s^{(5)}(x_1^2x_2^2x_3) = c_2c_3 - 3c_1c_4 + 5c_5.$$

§24. ANOTHER COMPUTATION

To learn how to avoid calculations like those which precede, we set

$$s(x_1^{\mu_1} \ . \ . \ . \ x_k^{\mu_k}) = \sum p_\nu c_1^{\nu_1} c_2^{\nu_2} \ . \ . \ . \ c_n^{\nu_n},$$

which we now know to be true. Replacing every x_i by λx_i, we change c_1 into λc_1, c_2 into $\lambda^2 c_2$, . . . , the exponent of λ depending on the degree of c_i in the x_i, and we obtain

$$\lambda^{\mu_1 + \cdots + \mu_k} s(x_1^{\mu_1} \ . \ . \ . \ x_k^{\mu_k}) = \sum \lambda^{\nu_1 + 2\nu_2 + \cdots + n\nu_n} p_\nu c_1^{\nu_1} c_2^{\nu_2} \ . \ . \ . \ c_n^{\nu_n}.$$

Hence we see that for every term of $\sum$ holds the relation

$$\nu_1 + 2\nu_2 + \ . \ . \ . \ + n\nu_n = \mu_1 + \ . \ . \ . \ + \mu_k.$$

The product of exponent and subscript of a letter is called the **weight of a letter.** The sum of the weights of letters in a term is called the **weight of a term** in those letters and is denoted by W. The sum of the degrees of letters in a term is called the **total degree** of a term in those letters and is denoted by D.

Since the weight in the c_i of all terms in $\sum$ thus is constant and equal to the total degree of s in the x_i, we can express s in terms of the c_i by simply restricting our choice of combinations $c_1^{\nu_1} \ldots c_n^{\nu_n}$ to such as satisfy the condition of weight. If then μ_1 is the highest degree of any x_i in a term of s, the total degree of the c_i in any term of $\sum$ must not exceed μ_1, which may exclude some combinations of the c_i that pass the restriction of weight.

Applying this we find:

(a) $s(x_1^2x_2^2) = \sum p_\nu c_1^{\nu_1} \ldots c_n^{\nu_n}$.

$W = 4$ permits $c_1^4,\ c_2^2,\ c_1c_3,\ c_4$

$D \leqq 2$ excludes c_1^4

$$s(x_1^2x_2^2) = c_2^2 + pc_1c_3 + qc_4.$$

(b) $s(x_1^2x_2^2x_3) = \sum p_\nu c_1^{\nu_1} \ldots c_n^{\nu_n}$.

$W = 5$ permits $c_1c_2^2,\ c_2c_3,\ c_1c_4,\ c_5$

$D \leqq 2$ excludes $c_1c_2^2$

$$s(x_1^2x_2^2x_3) = c_2c_3 + pc_1c_4 + qc_5.$$

For computation of the numerical coefficients, which remains to be done and which is done best by the use of special equations, we refer to the examples of chapter eight.

Also, there exist tables with the results of such computation. They are arranged according to the total degree of the symmetric sums, and we reproduce the table for the total degree four:

	c_4	c_3c_1	c_2^2	$c_2c_1^2$	c_1^4
$s(x_1^4)$	-4	4	2	-4	1
$s(x_1^3x_2)$	4	-1	-2	1	
$s(x_1^2x_2^2)$	2	-2	1		
$s(x_1^2x_2x_3)$	-4	1			
$s(x_1x_2x_3x_4)$	1				

This table gives, for instance:

$$s^{(4)}(x_1{}^4) = c_1{}^4 - 4c_1{}^2c_2 + 2c_2{}^2 + 4c_1c_3 - 4c_4$$
$$s^{(2)}(x_1{}^4) = c_1{}^4 - 4c_1{}^2c_2 + 2c_2{}^2.$$

Since symmetric functions are composed of symmetric sums, the rules of computation may be applied to them directly, the highest weight and degree setting the mark.

§25. RESULTANT[1]

Among the symmetric functions in the letters x_i which represent the roots of the general equation

$$f(x) = a_0x^m + a_1x^{m-1} + \cdots + a_m = 0$$

are the resultant and the discriminant.

If besides the function $f(x)$ we have another such function

$$g(x) = b_0x^n + b_1x^{n-1} + \cdots + b_n,$$

it will be convenient to denote the roots of $f(x)$ by α_i and the roots of $g(x)$ by β_i.

The two functions $f(x)$ and $g(x)$ have a root α_i in common if the product

$$g(\alpha_1)g(\alpha_2) \cdots g(\alpha_m)$$

vanishes with $g(\alpha_i)$ becoming zero. This product is evidently an integral function in the b_i. Since a permutation between the α_i only interchanges its factors, the product is a symmetric function of the α_i and as such by proposition (35) rationally expressible in terms of

$$\frac{a_1}{a_0}, \frac{a_2}{a_0}, \cdots, \frac{a_m}{a_0}.$$

Since the product is of degree not more than n in any one α_i, it is of total degree not more than n in the a_i/a_0. Multiplying it by $a_0{}^n$, we therefore obtain a function

$$R(f,g) = a_0{}^n g(\alpha_1)g(\alpha_2) \cdots g(\alpha_m)$$

which is integral in the a_i as it is in the b_i, and this function is called the **resultant** of $f(x)$ and $g(x)$.

[1] The rest of this chapter may be omitted on first reading.

It appears that

(36) **the vanishing resultant of two functions indicates that the two functions have a root in common.**

Since

$$g(x) = b_0(x - \beta_1)(x - \beta_2) \ . \ . \ . \ (x - \beta_n)$$

and hence

$$g(\alpha_1) = b_0(\alpha_1 - \beta_1)(\alpha_1 - \beta_2) \ . \ . \ . \ (\alpha_1 - \beta_n)$$
$$g(\alpha_2) = b_0(\alpha_2 - \beta_1)(\alpha_2 - \beta_2) \ . \ . \ . \ (\alpha_2 - \beta_n)$$
$$. \ . \ . \ . \ . \ . \ . \ . \ ,$$

the resultant takes the form

$$R(f,g) = a_0{}^n b_0{}^m \prod_{i,k} (\alpha_i - \beta_k), \quad [i = 1, \ . \ . \ . \ , m; k = 1, \ . \ . \ . \ , n$$

and a mere glance suffices to verify that its vanishing is a necessary and sufficient condition for any common root.

Interchanging the two functions $f(x)$ and $g(x)$ we may or may not alter the sign of the resultant, for

$$R(f,g) = (-1)^{mn} R(g,f)$$
$$= (-1)^{mn} b^m f(\beta_1) f(\beta_2) \ . \ . \ . \ f(\beta_n).$$

§26. RESULTANT AS DETERMINANT

To express the resultant of the two functions $f(x)$ and $g(x)$ in terms of their coefficients, we search for it anew. We already know that it is an integral function of the coefficients a_i and b_i.

Suppose that α_i is a common root of the two functions. Then we may set

$$f(x) = (x - \alpha_i) \ . \ f_1(x)$$

and

$$g(x) = (x - \alpha_i) \ . \ g_1(x),$$

where $f_1(x)$ is of degree $m - 1$ and $g_1(x)$ of degree $n - 1$ and both are integral functions of x. Eliminating $x - \alpha_i$ from the two equations, we find

$$f(x) \ . \ g_1(x) - f_1(x) \ . \ g(x) = 0.$$

Assuming that

$$f_1(x) = p_0 x^{m-1} + p_1 x^{m-2} + \ . \ . \ . \ + p_{m-2} x + p_{m-1}$$
$$g_1(x) = q_0 x^{n-1} + q_1 x^{n-2} + \ . \ . \ . \ + q_{n-2} x + q_{n-1}$$

and substituting, we find

$$(a_0q_0 - b_0p_0)x^{m+n-1} + (a_1q_0 + a_0q_1 - b_1p_0 - b_0p_1)x^{m+n-2} + \dots \\ + (a_mq_{n-2} + a_{m-1}q_{n-1} - b_np_{m-2} - b_{n-1}p_{m-1})x \\ + (a_mq_{n-1} - b_np_{m-1}) = 0.$$

This being an identity, the coefficients of x must be zero, whence p and q must satisfy the equations

$$\begin{aligned} a_0q_0 \qquad\qquad & -b_0p_0 & = 0 \\ a_1q_0 + a_0q_1 \qquad & - b_1p_0 - b_0p_1 & = 0 \\ & \dots\dots\dots & \\ a_mq_{n-2} + a_{m-1}q_{n-1} & \qquad -b_np_{m-2} - b_{n-1}p_{m-1} & = 0 \\ a_mq_{n-1} & \qquad - b_np_{m-1} & = 0 \end{aligned}$$

They will satisfy these equations if the determinant of their coefficients vanishes. With rows and columns interchanged, the determinant is

$$R_d = \left|\begin{array}{cccccccc} a_0 & a_1 & \dots & a_m & 0 & \dots & 0 \\ 0 & a_0 & \dots & a_{m-1} & a_m & \dots & 0 \\ & & \dots & \dots & \dots & & \\ 0 & 0 & \dots & \dots & \dots & \dots & a_m \\ b_0 & b_1 & \dots & \dots & \dots & \dots & 0 \\ 0 & b_0 & \dots & \dots & \dots & \dots & 0 \\ & & \dots & \dots & \dots & & \\ 0 & 0 & \dots & \dots & \dots & \dots & b_n \end{array}\right| \begin{array}{l} \left.\vphantom{\begin{array}{c}a\\a\\a\\a\end{array}}\right\} n \text{ rows} \\ \left.\vphantom{\begin{array}{c}a\\a\\a\\a\end{array}}\right\} m \text{ rows} \end{array}$$

This determinant is obviously homogeneous of total degree n in the a_i and of total degree m in the b_i, so that its terms are of the type

$$k \,.\, a_0^{\mu_0}a_1^{\mu_1}a_2^{\mu_2} \dots b_0^{\nu_0}b_1^{\nu_1}b_2^{\nu_2} \dots,$$

where

$$\mu_0 + \mu_1 + \mu_2 + \dots = n$$
$$\nu_0 + \nu_1 + \nu_2 + \dots = m.$$

These terms may be written in the form

$$k \,.\, a_0^n b_0^m \,.\, \left(\frac{a_1}{a_0}\right)^{\mu_1}\left(\frac{a_2}{a_0}\right)^{\mu_2} \dots \left(\frac{b_1}{b_0}\right)^{\nu_1}\left(\frac{b_2}{b_0}\right)^{\nu_2} \dots,$$

hence the determinant in the form

$$R_d = a_0^n b_0^m \,.\, \varphi(\alpha,\beta),$$

where φ is an integral function of the α_i and β_i.

As the determinant vanishes for any common root

$$\alpha_i = \beta_k,$$

the function φ is divisible by any $\alpha_i - \beta_k$, hence the determinant divisible by

$$a_0{}^n b_0{}^m \prod_{i,k} (\alpha_i - \beta_k). \quad [i = 1, \ldots, m; k = 1, \ldots, n$$

To find the quotient, we set

$$\begin{aligned} a_0{}^n b_0{}^m \prod_{i,k} (\alpha_i - \beta_k) &= a_0{}^n \prod_i [b_0(\alpha_i - \beta_1)(\alpha_i - \beta_2) \ldots (\alpha_i - \beta_n)] \\ &= a_0{}^n \prod_i g(\alpha_i) \qquad [i = 1, \ldots, m \\ &= (-1)^{mn} b_0{}^m \prod_k [a_0(\beta_k - \alpha_1) \ldots (\beta_k - \alpha_m)] \\ &= (-1)^{mn} b_0{}^m \prod_k f(\beta_k). \qquad [k = 1, \ldots, n \end{aligned}$$

Since $\prod_i g(\alpha_i)$ is homogeneous of total degree m in the b_i and $\prod_k f(\beta_k)$ is homogeneous of total degree n in the a_i, it follows that $a_0{}^n b_0{}^m \prod_{i,k} (\alpha_i - \beta_k)$ is homogeneous of total degree n in the a_i and of total degree m in the b_i. But so is the determinant, and their quotient can be only numerical. Comparing the leading term of the determinant, which is $a_0{}^n b_n{}^m$, with the corresponding term of $a_0{}^n b_0{}^m \prod_{i,k} (\alpha_i - \beta_k)$, which is

$$\begin{aligned} &a_0{}^n [b_0(-\beta_1)(-\beta_2) \ldots (-\beta_n)]^m \\ &= a_0{}^n [b_0(-1)^n \beta_1 \beta_2 \ldots \beta_n]^m \\ &= a_0{}^n \left[b_0 \frac{b_n}{b_0} \right]^m \\ &= a_0{}^n b_n{}^m, \end{aligned}$$

we find that the quotient is 1, so that

$$R_d = a_0{}^n b_0{}^m \prod_{i,k} (\alpha_i - \beta_k) = R(f,g)$$

and the determinant appears to be the resultant itself expressed as function of the coefficients a_i and b_i.

The resultant, expressed as function of the roots α_i and β_i, evidently is homogeneous of total degree mn in these roots. Considering that a_k is of degree k in the α_i and b_k is of degree k in the β_i, we notice that the terms of the resultant expressed as function of the coefficients a_i and b_i are of constant weight in these coefficients, for

$$\mu_1 + 2\mu_2 + 3\mu_3 + \dots$$
$$+ \nu_1 + 2\nu_2 + 3\nu_3 + \dots = mn.$$

Hence we have the proposition·

(37) **The resultant of two functions is homogeneous of total degree *mn* in their roots and expressible as determinant of constant weight *mn* in their coefficients, if *m* and *n* are the degrees of the functions.**

§27. DISCRIMINANT

If $f'(x)$ denotes the derivative of

$$f(x) = a_0x^m + a_1x^{m-1} + \dots + a_m,$$

$R(f,f')$ is divisible by a_0 as we see from its determinant form which permits a_0 to be taken out of the first column. We set

$$\frac{1}{a_o}R(f,f') = l \,.\, D(f),$$

introducing the coefficient l for an adjustment to be made presently, and call D the **discriminant** of the function $f(x)$.

Since

$$R(f,f') = a_0{}^{m-1}f'(\alpha_1)f'(\alpha_2) \dots f'(\alpha_m),$$

we have

$$l \,.\, D(f) = a_0{}^{m-2}\prod_i f'(\alpha_i). \qquad [i = 1, \dots, m$$

But the derivative of $f(x)$ for α_i is

$$f'(\alpha_i) = a_0[(\alpha_i - \alpha_1)(\alpha_i - \alpha_2) \dots (\alpha_i - \alpha_{i-1})(\alpha_i - \alpha_{i+1}) \dots (\alpha_i - \alpha_m)].$$

This gives

$$\prod_i f'(\alpha_i) = (-1)^{\frac{m(m-1)}{2}} a_0{}^m \prod_{i \lessgtr k} (\alpha_i - \alpha_k)^2, \; [i, k = 1, \dots, m$$

with a coefficient taking care of the sign because there are $m(m-1)$ factors $\alpha_i - \alpha_k$ and every such factor appears twice with opposite sign in the left member while appearing twice with the same sign in the right member.

Substituting into the expression for D, we obtain

$$l \,.\, D(f) = (-1)^{\frac{m(m-1)}{2}} a_0^{2m-2} \prod_{i \lessgtr k} (\alpha_i - \alpha_k)^2. \quad [i, k = 1, \ldots, m$$

Now we set

$$l = (-1)^{\frac{m(m-1)}{2}},$$

and our formulas become definitively

$$\frac{1}{a_0} R(f,f') = (-1)^{\frac{m(m-1)}{2}} D(f),$$

$$D(f) = a_0^{2m-2} \prod_{i \lessgtr k} (\alpha_i - \alpha_k)^2.$$

Since

$$R(f,f') = 0$$

means a common root of $f(x)$ and $f'(x)$, and by proposition (12) such a common root is a multiple root of $f(x)$, we conclude that

(38) **the discriminant**

$$\boxed{D(f) = (-1)^{\frac{m(m-1)}{2}} \frac{R(f,f')}{a_0}}$$

of a function $f(x)$ of degree m vanishes whenever that function has a multiple root.

For $a_0 = 1$ we obtain the non-homogeneous form of the discriminant, which we denote by Δ:

$$\Delta(f) = \prod_{i \lessgtr k} (\alpha_i - \alpha_k)^2.$$

In case of the quadratic equation

$$f(x) = a_0 x^2 + a_1 x + a_2 = 0,$$

the discriminant is

$$D = a_1^2 - 4a_0 a_2.$$

In case of the cubic equation

$$f(x) = a_0x^3 + a_1x^2 + a_2x + a_3 = 0,$$

the discriminant is

$$D = a_1^2a_2^2 + 18a_0a_1a_2a_3 - 4a_0a_2^3 - 4a_1^3a_3 - 27a_0^2a_3^2.$$

We may notice that already

$$\sqrt{\Delta} = \prod_{i \lessgtr k} (\alpha_i - \alpha_k)$$

disappears when two roots are alike; but it is not symmetric in the α_i and hence not rationally expressible in terms of the a_i.

We may also notice that it makes a difference with $\sqrt{\Delta}$ whether we set $i < k$ or $i > k$, while it makes no difference with Δ, for

$$\prod_{i<k} (\alpha_i - \alpha_k) = (-1)^{\frac{m(m-1)}{2}} \prod_{i>k} (\alpha_i - \alpha_k)$$

where

$$\begin{aligned} \prod_{i<k} (\alpha_i - \alpha_k) = (\alpha_1 - \alpha_2)(\alpha_1 - \alpha_3) \; . \; . \; . \; &(\alpha_1 - \alpha_m) \\ (\alpha_2 - \alpha_3) \; . \; . \; . \; &(\alpha_2 - \alpha_m) \\ . \quad . \quad . \quad . \quad . \quad . \quad & \end{aligned}$$

and

$$\begin{aligned} \prod_{i>k} (\alpha_i - \alpha_k) = (\alpha_2 - \alpha_1)(\alpha_3 - \alpha_1) \; . \; . \; . \; &(\alpha_m - \alpha_1) \\ (\alpha_3 - \alpha_2) \; . \; . \; . \; &(\alpha_m - \alpha_2) \\ . \quad . \quad . \quad . \quad . \quad . \quad & \end{aligned}$$

CHAPTER VI

COMPOSITION OF SYMMETRIC GROUP

§28. COMPOSITION-SERIES

A complete series of normal subgroups, each the greatest such group in the preceding, is called the **composition-series** of a group. The composition-series of the symmetric group begins with the symmetric group itself and ends with identity as every composition-series necessarily does.

The indices of the composition-series, each one taken as the index of a group in the group preceding it, are called the **factors of composition.**

To find a composition-series, it is useful to know that normal subgroups of a group G are composed of all permutations which are common to conjugate subgroups of G.

For suppose such conjugate subgroups of G are

$$H_1, H_2, \ldots, H_j,$$

The permutations which are common to these subgroups form a group, since with any two of them contained in H_i also their product is contained in H_i by the very definition of a group. And since the transforms of H_1 by permutations in G are the same conjugate subgroups

$$H_1, t_2^{-1}H_1t_2, \ldots, t_j^{-1}H_1t_j,$$

the group of common permutations, which is part of any one of them as it is of H_1, is identical with its transforms and hence a normal subgroup of G.

A group of all permutations which are common to conjugate subgroups is called their **greatest common subgroup**[1] and denoted by D, so that we have the proposition:

(39) **a greatest common subgroup is normal in the group.**[2]

The question may occur why there cannot be distinct conjugates D' and D'' such that D' of H_1 is transformed into D'' of

[1] Called "Durchschnitt" in German.

[2] Compare §54.

H_i while D' of H_i comes from D'' of H_1. But this implies the presence of both D' and D'' in any H_i and more common permutations than we have in fact.

Thus the conjugate subgroups in the example of §19 have the greatest common subgroup

$$D = 1, (12)(34), (13)(24), (14)(23)$$

which is normal in S, while

$$H_1 = 1, (12)(34)$$
$$H_2 = 1, (13)(24)$$
$$H_3 = 1, (14)(23)$$

are common subgroups but not normal since they are not greatest.

§29. ALTERNATING FUNCTION

The greatest possible subgroup of the symmetric group S on n letters x_i is one of index

$$j = 2$$

and order

$$r = \frac{n!}{2}.$$

It exists by proposition (20) because there exist two-valued functions of these x_i, which are called **alternating functions.**[1]

Suppose that A_i is an alternating function with no other values than A_1 and A_2. If the group of A_1 is

$$\{A\} = 1, s_2, \ldots, s_r$$

and the permutation t is not in it, then the permutations of

$$\{A\}t = t, s_2t, \ldots, s_rt$$

convert A_1 into A_2 and

$$\{S\} = \{A\} + \{A\}t.$$

Since both sets At and tA contain all permutations that are in S and not in A, we have

$$\{A\}t = t\{A\}$$

and

$$t^{-1}\{A\}t = \{A\}.$$

[1] Cf. §8.

Hence A is normal: it is the group of A_2 as it is the group of A_1, while the permutations of At convert A_2 into A_1 as they convert A_1 into A_2.

As a permutation of S either leaves the functions A_1 and A_2 unaltered or interchanges them, the sum

$$A_1 + A_2$$

is a symmetric function while the difference

$$A_1 - A_2 = \Phi$$

has the conjugate value

$$A_2 - A_1 = -\Phi$$

and is an alternating function.

Taking now a permutation that alters Φ, we resolve it into transpositions. As we successively apply these transpositions to Φ, we must strike one that changes the sign of Φ, for the permutation does so. If it is (x_1x_2), so that

$$\Phi(x_1, x_2, \ldots, x_n) = -\Phi(x_2, x_1, \ldots, x_n),$$

we set

$$x_1 = x_2;$$

this gives

$$\Phi(x_1, \ldots, x_n) = -\Phi(x_1, \ldots, x_n) = 0,$$

since both members differ by their sign alone.

It appears that Φ is divisible by $x_1 - x_2$. But then Φ^2 is divisible by $(x_1 - x_2)^2$ and, as it evidently is symmetric, by every $(x_i - x_k)^2$. Hence Φ in turn is divisible by every $x_i - x_k$ and equals

$$\sqrt{\Delta} = \prod_{i \lessgtr k} (x_i - x_k)$$

itself or multiplied by a symmetric function. Setting

$$A_1 + A_2 = 2S_1$$

$$\Phi = A_1 - A_2 = 2S_2\sqrt{\Delta},$$

we find that

(40) **the general form of alternating functions is**

$$\boxed{\begin{aligned} A_1 &= S_1 + S_2\sqrt{\Delta} \\ A_2 &= S_1 - S_2\sqrt{\Delta} \end{aligned}}.$$

With the alternating functions exists the group of these functions which is called the **alternating group** and is denoted by A or, to guard against confusion, by $\{A\}$:

(41) **The alternating group is a normal subgroup of index two in the symmetric group.**

It is the greatest subgroup of the symmetric group, containing one half of its permutations. The simplest function that belongs to it is the root $\sqrt{\Delta}$ of the discriminant.

§30. ALTERNATING GROUP

Every transposition alters the sign of $\sqrt{\Delta}$ and the value of an alternating function, as we can verify on any example. Writing subscripts only, we have in five letters x_i, for instance:

$$\begin{aligned} \sqrt{\Delta} = (1 - 2)(1 - 3)(1 - 4)(1 - 5) \\ (2 - 3)(2 - 4)(2 - 5) \\ (3 - 4)(3 - 5) \\ (4 - 5). \end{aligned}$$

By the transposition (24), say, the factors $(1 - 2)$ and $(2 - 5)$ change only place with $(1 - 4)$ and $(4 - 5)$, the factors $(2 - 3)$ and $(3 - 4)$ change place and also sign. While this leaves $\sqrt{\Delta}$ unaltered, there is one factor, the factor $(2 - 4)$ containing the numbers of the transposition, which changes its own sign and that of $\sqrt{\Delta}$.

Hence we conclude that any even number of transpositions leaves $\sqrt{\Delta}$ unaltered while any odd number of them alters its sign. If then some permutation once breaks up into an even number of transpositions, it always does so since it cannot alter and not alter the sign of $\sqrt{\Delta}$. Such a permutation is called an **even permutation**; while another which is formed by an odd number of transpositions is called **odd.**

(42) **A circular permutation, or the cycle of a non-circular permutation, is odd or even according as its degree is even or odd; a non-circular permutation is odd or even according as it contains an odd or even number of odd cycles.**

For instance:

$$(123) = (12)(13)$$

is even while

$$(1234) = (12)(13)(14)$$

is odd, and

$$(1234)(56)$$

is even because it contains two odd cycles. Two similar permutations evidently are both odd or both even.

It is clear that

(43) **the alternating group is composed of all even permutations in the symmetric group,**

all those that leave unaltered the sign of $\sqrt{\Delta}$. We readily admit that such permutations form a group, for the product of even permutations is again an even permutation; also that this group is normal, for the transform of an even permutation is by proposition (30) again even.

Every permutation on n letters x_i can by proposition (33) be represented as the product of transpositions which are in the set $(x_1x_i)_2^n$; every even permutation therefore as the product of such paired transpositions, which is to say as the product of circular permutations

$$(x_1x_ix_k) = (x_1x_i)(x_1x_k)$$

of order three. But we can easily verify that

$$(x_1x_ix_k) = (x_1x_2x_i)^2(x_1x_2x_k),$$

whence the permutations of the alternating group are resolvable into cycles of order three and, if we so choose, into such as give the $n - 2$ permutations

$$(x_1x_2x_i)_3^n = (x_1x_2x_3), (x_1x_2x_4), \ldots, (x_1x_2x_n).$$

It follows that

(44) **the alternating group on n letters x_i is generated by the $n - 2$ independent permutations $(x_1x_2x_i)_3^n$:**

$$\boxed{A^n = \{(x_1x_2x_i)_3^n\}};$$

and a group including these permutations is the alternating if not the symmetric group.

For the lowest degrees the alternating groups are:

$A^2 = 1$

$A^3 = 1, (123), (132)$

$A^4 =$

1	(12)(34)	(13)(24)	(14)(23)
(123) (132)	(243) (143)	(142) (234)	(134) (124)

If we denote the numbers

$$1, 2, 3, 4, \ldots$$

in any order whatever by

$$i_1, i_2, i_3, i_4, \ldots$$

we can assert that the two permutations

$$t_1 = \begin{pmatrix} 1234 & \ldots \\ i_1 i_2 i_3 i_4 & \ldots \end{pmatrix}$$

and

$$t_2 = \begin{pmatrix} 1234 & \ldots \\ i_2 i_1 i_3 i_4 & \ldots \end{pmatrix}$$

cannot both be odd or even, for their product t_1t_2 is odd since the product

$$t_1t_2^{-1} = (12)$$

is odd and the permutations t_2 and t_2^{-1} are similar by proposition (18).

Hence either t_1 or t_2 is in the alternating group, and

(45) **a normal subgroup of the alternating group contains with one circular permutation of order three every possible such permutation.**

For suppose it contains the permutation

$$s = (123).$$

If the permutation t_1 is in the alternating group, its normal subgroup contains the permutation

$$t_1^{-1}st_1 = (i_1i_2i_3);$$

and if the permutation t_2 is in the alternating group, its normal subgroup contains the permutation

$$t_2^{-1}s^2t_2 = (i_1i_2i_3).$$

Thus it cannot help containing any circular permutation $(i_1i_2i_3)$ of order three.

§31. COMPOSITION OF S AND A

So far we know about the composition-series of the symmetric group on n letters x_i that it is

$$S \quad A \; . \; . \; . \; 1.$$

We now ask the question that could make our hearts beat faster if we knew its significance: Are there in the symmetric group normal subgroups other than the alternating group? And are there normal subgroups in the alternating group?

Supposing there is in either of these groups a normal subgroup N, so that we have

$$t_i^{-1}Nt_i = N$$

at least for every permutation t_i of the alternating group, we search for those permutations of N that have the lowest degree. Of course, we shall disregard identity and not count toward the degree letters that remain fixed.

The permutations of lowest degree cannot be circular of degree more than three, nor non-circular containing a cycle of degree more than three, for a permutation

$$s_1 = (1234 \; . \; . \; . \;) \; . \; . \; .$$

of N is transformed by the permutation

$$t = (123)$$

of A into the permutation

$$t^{-1}s_1t = (2314 \; . \; . \; . \;) \; . \; . \; . = s_2$$

of N. Hence N contains also the permutation

$$s_1s_2^{-1} = (2)(31 \; . \; . \; . \;) \; . \; . \; .$$

which is of lower degree than s_1 since the letter with subscript 2 is excluded and no letter added.

The permutations of lowest degree cannot be non-circular with a cycle of degree three, for a permutation

$$s_1 = (123)(4 \; . \; . \; . \;) \; . \; . \; .$$

of N is transformed by the permutation

$$t = (234)$$

of A into the permutation

$$t^{-1}s_1t = (134)(2\ \ldots\ )\ \ldots = s_2$$

of N. Hence N contains also the permutation

$$s_1s_2 = (3)(24\ \ldots\ )\ \ldots$$

which is of lower degree than s_1.

Circular permutations of degree three qualify as permutations of lowest degree in N. But containing one of them, N contains them all by proposition (45) and is by proposition (44) the alternating if not the symmetric group.

Likewise single transpositions qualify; but containing such, N is a subgroup of the symmetric and not the alternating group, contains them all by proposition (31) and is by proposition (33) the symmetric group itself.

Non-circular permutations of degree more than four with cycles of degree two are impossible, for a permutation

$$s_1 = (12)(34)\ \ldots$$

of N is transformed by the permutation

$$t = (345)$$

of A into the permutation

$$t^{-1}s_1t = (12)(45)\ \ldots = s_2$$

of N. Hence N contains also the permutation

$$s_1s_2 = (1)(2)(35\ \ldots\ )\ \ldots$$

which lost two letters although it may have gained one and is of lower degree as compared with s_1.

This conclusion is not valid, however, for a group of degree four. The two possible transforms of

$$s_1 = (12)(34)$$

then are

$$s_2 = (13)(24)$$
$$s_3 = (14)(23),$$

and these permutations together with identity constitute the **quadratic group**[1]

$$V = 1, (12)(34), (13)(24), (14)(23)$$

of degree and order four which is normal in both the symmetric and alternating groups of degree four.

The disappointing result of our investigation is this:

(46) **The symmetric group has no normal subgroup other than the alternating group, and the alternating group has no normal subgroup at all. Only the symmetric and alternating groups of degree four are exceptions.**

Hence the symmetric group is always composite; but the alternating group is always simple, except when its degree is four.

The quadratic group is unique in many ways; although not circular, it is composed of commutative permutations: for instance,

$$(12)(34) . (13)(24) = (13)(24) . (12)(34) = (14)(23).$$

As a six-valued function belonging to V we mention

$$\psi = (x_1 - x_2)(x_3 - x_4);$$

another example occurs in §39. But not every six-valued function belongs to V; thus the function

$$\xi = x_1 + x_2 - x_3 - x_4$$

belongs to the eroup

$$W = 1, (12), (34), (12)(34).$$

§32. SUBGROUPS OF S AND A

If the symmetric group has no other normal subgroup than the alternating group, it has other subgroups that are not normal. But

(47) **the symmetric group on n letters x_i has no subgroup of index between 2 and n, except when $n = 4$.**

This implies that a function in n letters x_i which has fewer than n values cannot have more than 2 values, unless in the unruly case of $n = 4$.

[1] Called "Vierergruppe" in German.

For suppose that a function ψ_1 takes under the symmetric group S fewer than n conjugate values, and suppose that these values are

$$\psi_1, \psi_2, \ldots, \psi_j$$

belonging to conjugate subgroups [$j < n$

$$H_1, H_2, \ldots, H_j$$

of S. The $n!$ permutations on the x_i which are in S can do no more than interchange the ψ_i, as it was explained in §16, and interchange them in $j!$ possible ways only.

Since $j!$ is less than $n!$, there must be distinct permutations on the x_i that operate the same permutation on the ψ_i. Let t and u be two of them, then

$$s = tu^{-1}$$

is a permutation different from identity that leaves the ψ_i unaltered. But permutations which do so are contained in every H_i; they are the only common permutations of the H_i and compose by proposition (39) a normal subgroup

$$D = 1, s_2, \ldots, s_r$$

of the symmetric group S.

By proposition (46), this normal subgroup D can be the alternating group alone, for it is not identity. Containing then the alternating group, the subgroup H_1 is identical with it if not symmetric; and belonging to H_1, the function ψ_1 is an alternating if not a symmetric function. Having less than n values, it cannot have more than two values, which proves the proposition.

Likewise, the number $n!/2$ of permutations in the alternating group A is greater than $j!$ for j less than n,[1] if a function ψ_1 takes j values ψ_i under A, and there must be distinct permutations in A which operate the same permutation on the ψ_i.

Since a subgroup of index between 1 and n in the alternating group then calls for a normal subgroup between the alternating group and identity, and such a subgroup by proposition (46) does not exist, it appears that

(48) **the alternating group on n letters x_i has no subgroup of index between 1 and n, except when $n = 4$.**

[1] This is meant for $n > 2$.

The subgroup of index n in the symmetric group is therefore not a subgroup of the alternating group, as we shall observe presently when identifying such a subgroup.

Another way of putting the proof of our propositions follows. The number ρ of permutations between the ψ_i under S is

$$\rho = n!/r,$$

while necessarily

$$j! \geqq \rho.$$

As $j > 2$ means[1]

$$r = 1,$$

we have

$$j \geqq n.$$

An n-valued function ψ_1 in n letters x_i is readily constructed if we set ψ_1 equal to x_1 or to the sum of all x_i except x_1. Thus we have for $n = 3$ the three-valued functions

$$\psi_1 = x_1, \qquad \psi_2 = x_2, \qquad \psi_3 = x_3,$$

or

$$\psi_1 = x_2 + x_3, \quad \psi_2 = x_1 + x_3, \quad \psi_3 = x_1 + x_2.$$

This verifies that a subgroup of index n in S^n is S^{n-1}, the group for instance of ψ_1 containing all permutations which act upon the x_i other than x_1.

In the exceptional case of four letters x_i there is between the alternating group and identity a normal subgroup of order

$$r = 4$$

in the symmetric group. Hence we have

$$j \geqq 3$$

and look for a function with

$$2 < j = 3 < 4.$$

One such function is

$$\psi_1 = x_1x_2 + x_3x_4,$$

which occurred in §17.

The number ρ of permutations between the ψ_i under A is

$$\rho = \frac{n!}{2}/r,$$

while again

$$j! \geqq \rho.$$

As $j > 1$ means[1]

$$r = 1,$$

[1] By proposition (46).

we have

$$j! \geqq \frac{n!}{2}.$$

The exceptional case

$$r = 4$$

when $n = 4$ gives

$$j! \geqq 6$$

and permits

$$1 < j = 3 < 4.$$

The same function

$$\psi_1 = x_1x_2 + x_3x_4$$

will serve as an example.

We note that our results may also be expressed by saying that symmetric and alternating functions exist for any number of letters; three-valued functions exist in three and four letters only; and after that no n-valued function can be constructed in more than n letters.

§33. GROUP ON FUNCTIONS

We noticed that under the partitions of a group

$$G = N + Nt_2 + \ . \ . \ . + Nt_j$$

with respect to a normal subgroup N conjugate values of a function interchange in as many ways as there are partitions if they belong to N or to groups[1] containing N as greatest common subgroup.

We have to add that the permutations between such conjugate values of a function, as effected by the permutations of G, form a group, which we express in the proposition:

(49) **The permutations under G between conjugate functions ψ_i belonging to a normal subgroup N of G or to groups containing N as greatest common subgroup compose a group Γ.**

For suppose that these permutations are

$$1, \tau_2, \ . \ . \ . \ , \tau_j$$

interchanging the conjugate values

$$\psi_1, \psi_2, \ . \ . \ . \ , \psi_j$$

[1] Whether subgroups of G or not.

as permutations in the partitions

$$N, Nt_2, \ldots, Nt_j$$

of G do. Set

$$\tau_a = \begin{pmatrix} \psi_1 \psi_2 \ldots \psi_j \\ \psi_{1a}\psi_{2a} \ldots \psi_{ja} \end{pmatrix} = \begin{pmatrix} \psi_i \\ \psi_{ia} \end{pmatrix}$$

and

$$\tau_b = \begin{pmatrix} \psi_i \\ \psi_{ib} \end{pmatrix} = \begin{pmatrix} \psi_{ia} \\ \psi_{iab} \end{pmatrix},$$

which we can do because the ψ_{ia} represent in some order or other all the ψ_i. Then we have

$$\tau_a\tau_b = \begin{pmatrix} \psi_i \\ \psi_{iab} \end{pmatrix};$$

and since the permutation

$$t_at_b = t_c$$

is contained in G, the permutation

$$\tau_a\tau_b = \tau_c$$

is contained among the τ_i which consequently compose a group

$$\Gamma = 1, \tau_2, \ldots, \tau_j.$$

This we illustrate on the permutations between the functions

$$\psi_1 = x_1x_2 + x_3x_4$$
$$\psi_2 = x_1x_3 + x_2x_4$$
$$\psi_3 = x_1x_4 + x_2x_3$$

conjugate under the symmetric group on the x_i:

Γ on the ψ_i		S on the x_i	
1	$\psi_1\psi_2\psi_3$	1, (12)(34), (13)(24), (14)(23) $= N$	
$\tau_2 = (123)$	$\psi_2\psi_3\psi_1$	(234), (132), (143), (124) $= Nt_2$	$t_2 = (234)$
$\tau_3 = (132)$	$\psi_3\psi_1\psi_2$	(243), (142), (123), (134) $= Nt_3$	$t_3 = (243)$
$\tau_4 = (12)$	$\psi_2\psi_1\psi_3$	(23), (1342), (1243), (14) $= Nt_4$	$t_4 = (23)$
$\tau_5 = (13)$	$\psi_3\psi_2\psi_1$	(24), (1432), (13), (1234) $= Nt_5$	$t_5 = (24)$
$\tau_6 = (23)$	$\psi_1\psi_3\psi_2$	(34), (12), (1423), (1324) $= Nt_6$	$t_6 = (34)$

CHAPTER VII

THEORY OF LAGRANGE

§34. RESOLVENT EQUATION

In the endeavor to find a solution of the general equation

$$f(x) = x^n - c_1x^{n-1} + \ . \ . \ . \ \pm c = 0$$

we now turn to functions of its roots

$$x_1, x_2, \ . \ . \ . \ , x_n,$$

for we know from chapter two that the computation of such functions may imply the solution of the general equation.

Any such function takes a certain number of conjugate values under every group containing the group of the function. An equation whose roots such conjugate values of a function are is called a **resolvent equation or resolvent** of the general equation, and the following proposition applies:

(50) **The conjugate values which a function takes under a group to whose subgroup the function belongs are roots of a resolvent equation whose degree equals the index of the subgroup in the group and whose coefficients are not altered by the group.**

For suppose that some function ψ_1 of the x_i belongs to a subgroup H of index j in the group G on the x_i. Its conjugate values

$$\psi_1, \psi_2, \ . \ . \ . \ , \psi_j$$

under G are roots of the resolvent equation

$$r(\psi) = (\psi - \psi_1)(\psi - \psi_2) \ . \ . \ . \ (\psi - \psi_j) = 0$$

of degree j with coefficients

$$\psi_1 + \psi_2 + \psi_3 + \ . \ . \ . \ + \psi_j$$
$$\psi_1\psi_2 + \psi_1\psi_3 + \ . \ . \ . \ + \psi_{j-1}\psi_j$$
$$. \ . \ . \ . \ . \ . \ .$$
$$\psi_1\psi_2\psi_3 \ . \ . \ . \ \psi_j$$

which are symmetric in the ψ_i. Since a permutation of G can do no more than interchange the ψ_i, as explained in §16, it leaves the coefficients of the resolvent equation unaltered, and the proposition follows.

In this connection we say that $r(\psi) = 0$ is a resolvent equation for H under G.

§35. LAGRANGE'S THEOREM

That the solution of a resolvent equation may imply the solution of the general equation follows from a proposition known as **Lagrange's Theorem**[1] and published in a celebrated contribution of Lagrange to the Memoirs of the Academy of Berlin for 1770–71:

(51) **If a rational function φ_1 in the roots x_i of the general equation remains unaltered by all those permutations on the x_i that leave another rational function ψ_1 of the x_i unaltered, then the function φ_1 is rationally expressible in terms of the function ψ_1 and the coefficients of the general equation.**

This is to say that the function φ_1 can be computed from the function ψ_1 and the coefficients of the general equation by rational operations. We note incidentally that by the conditions of the theorem the function ψ_1 belongs to the group of the function φ_1 or to one of its subgroups.

Let φ_1 and ψ_1 be two rational functions of the x_i belonging first to the same group H of index j in the symmetric group S on the x_i. We shall prove that φ_1 is rationally expressible in terms of ψ_1 and the coefficients c_i of the general equation.

The conjugate values

$$\varphi_1, \varphi_2, \ldots, \varphi_j$$
$$\psi_1, \psi_2, \ldots, \psi_j$$

of φ_1 and ψ_1 under S can be arranged so that permutations of S converting φ_i into φ_k also convert ψ_i into ψ_k. The function

$$\psi_1{}^k\varphi_1 + \psi_2{}^k\varphi_2 + \ldots + \psi_j{}^k\varphi_j$$

[1] Compare the statement of Lagrange's Theorem in §71. Another proof is given in §69. Lagrange lived 1736–1813.

then is symmetric in the x_i since no permutation on the x_i can do more than interchange its terms. Consequently it is rational in the c_i by proposition (35), and we have a system

$$\begin{aligned}
\varphi_1 + \varphi_2 + \dots + \varphi_j &= r_0(c_i) \\
\psi_1\varphi_1 + \psi_2\varphi_2 + \dots + \psi_j\varphi_j &= r_1(c_i) \\
\psi_1^2\varphi_1 + \psi_2^2\varphi_2 + \dots + \psi_j^2\varphi_j &= r_2(c_i) \\
\dots\dots\dots \\
\psi_1^{j-1}\varphi_1 + \psi_2^{j-1}\varphi_2 + \dots + \psi_j^{j-1}\varphi_j &= r_{j-1}(c_i)
\end{aligned}$$

of j equations linear in the φ_i. We need no more equations to solve for φ_1; and we can have no more since powers of ψ_i higher than $j - 1$ can be eliminated by rational operations. For ψ_i satisfies by proposition (50) an equation

$$r(\psi) = \psi^j + A_1\psi^{j-1} + A_2\psi^{j-2} + \dots = 0$$

with coefficients that are symmetric in the x_i and as such rational in the c_i, whence

$$\begin{aligned}
\psi_i^j &= -A_1\psi_i^{j-1} - A_2\psi_i^{j-2} - \dots \\
\psi_i^{j+1} &= -A_1\psi_i^j - A_2\psi_i^{j-1} - \dots \\
&= (A_1^2 - A_2)\psi_i^{j-1} + (A_1A_2 - A_3)\psi_i^{j-2} + \dots \\
&\dots\dots\dots
\end{aligned}$$

Solving the system of equations in the φ_i given above for φ_1, we obtain

$$\varphi_1 = \frac{\begin{vmatrix} r_0 & 1 & 1 & \dots & 1 \\ r_1 & \psi_2 & \psi_3 & \dots & \psi_j \\ r_2 & \psi_2^2 & \psi_3^2 & \dots & \psi_j^2 \\ \cdot & \cdot & \cdot & \cdot & \cdot \\ r_{j-1} & \psi_2^{j-1} & \psi_3^{j-1} & \dots & \psi_j^{j-1} \end{vmatrix}}{\begin{vmatrix} 1 & 1 & 1 & \dots & 1 \\ \psi_1 & \psi_2 & \psi_3 & \dots & \psi_j \\ \psi_1^2 & \psi_2^2 & \psi_3^2 & \dots & \psi_j^2 \\ \cdot & \cdot & \cdot & \cdot & \cdot \\ \psi_1^{j-1} & \psi_2^{j-1} & \psi_3^{j-1} & \dots & \psi_j^{j-1} \end{vmatrix}}$$

Since the determinant of the denominator vanishes for every

$$\psi_i = \psi_k, \qquad [i \neq k$$

this determinant is divisible by every $\psi_i - \psi_k$ and hence by

$$\prod(\psi_i - \psi_k) = \sqrt{\Delta_\psi},$$

where we take

$$i > k.$$

Having j values ψ_i, we can pick $\psi_i - \psi_k$ such that $i > k$ in $j(j-1)/2$ different ways, whence $\prod(\psi_i - \psi_k)$ is homogeneous of total degree $j(j-1)/2$ in the ψ_i. But so is the determinant as

$$1 + 2 + \ldots + (j-1) = \frac{j(j-1)}{2},$$

and their quotient can be only numerical. From the leading term of the determinant, as compared with the corresponding term of $\prod(\psi_i - \psi_k)$, it is seen to be 1, whence

$$\begin{vmatrix} 1 & 1 & 1 & \ldots & 1 \\ \psi_1 & \psi_2 & \psi_3 & \ldots & \psi_j \\ \psi_1^2 & \psi_2^2 & \psi_3^2 & \ldots & \psi_j^2 \\ \cdot & \cdot & \cdot & \cdot & \cdot \\ \psi_1^{j-1} & \psi_2^{j-1} & \psi_3^{j-1} & \ldots & \psi_j^{j-1} \end{vmatrix} = \sqrt{\Delta_\psi}$$

Denoting now by T the determinant of the numerator, we may set

$$\varphi_1 = \frac{T}{\sqrt{\Delta_\psi}} = \frac{T \cdot \sqrt{\Delta_\psi}}{\Delta_\psi},$$

where Δ_ψ is symmetric in the ψ_i and hence the x_i and as such rational in the c_i. To complete our proof, it remains to investigate the numerator

$$T \cdot \sqrt{\Delta_\psi} = \begin{vmatrix} r_0 & 1 & 1 & \ldots & 1 \\ r_1 & \psi_2 & \psi_3 & \ldots & \psi_j \\ r_2 & \psi_2^2 & \psi_3^2 & \ldots & \psi_j^2 \\ \cdot & \cdot & \cdot & \cdot & \cdot \\ r_{j-1} & \psi_2^{j-1} & \psi_3^{j-1} & \ldots & \psi_j^{j-1} \end{vmatrix} \cdot \sqrt{\Delta_\psi}$$

It is a function symmetric in the conjugate values

$$\psi_2, \psi_3, \ldots, \psi_j$$

other than ψ_1, for interchanging any two such values ψ_i we alter the sign of both T and $\sqrt{\Delta_\psi}$ but do not alter their product. To examine such a function, we take the equation

$$(\psi - \psi_1)(\psi - \psi_2) \ldots (\psi - \psi_j)$$
$$= \psi^j + A_1\psi^{j-1} + A_2\psi^{j-2} + \ldots + A_j = 0$$

with coefficients rational in the c_i and dividing out $\psi - \psi_1$ obtain the equation

$$\begin{aligned}(\psi - \psi_2) \dots (\psi - \psi_j) \\ = \psi^{j-1} + (\psi_1 + A_1)\psi^{j-2} + (\psi_1^2 + A_1\psi_1 + A_2)\psi^{j-3} + \dots \\ = \psi^{j-1} + B_1\psi^{j-2} + B_2\psi^{j-3} + \dots = 0,\end{aligned}$$

where

$$\begin{aligned}B_1 &= \psi_1 + A_1 \\ B_2 &= \psi_1^2 + A_1\psi_1 + A_2 \\ &\dots\dots\end{aligned}$$

Hence it appears that a symmetric function of the roots ψ_i other than ψ_1, rationally expressible in terms of the elementary symmetric functions B_i of those roots, is expressible so in terms of ψ_1 and the A_i, and therefore expressible so in terms of ψ_1 and the c_i:

$$\text{Sym.}\ (\psi_2, \dots, \psi_j) = \text{Rat.}\ (\psi_1, c_i).$$

Since the numerator in the expression for φ_1 is such a function and its coefficients are rational with the r_i, we may set

$$\varphi_1 = \frac{\text{Rat.}\ (\psi_1, c_i)}{\Delta_\psi}$$

or

$$\varphi_1 = R(\psi_1, c_i)$$

with R integral in ψ_1.

This proves Lagrange's Theorem for functions belonging to the same group. Inasmuch as the coefficients c_i of the general equation are to be regarded as rational, we may set also

$$\varphi_1 = R(\psi_1).$$

Since powers of ψ_1 higher than $j - 1$ can be eliminated, the expression for φ_1 can be reduced to the form

$$\varphi_1 = \frac{R_1\psi_1^{j-1} + R_2\psi_1^{j-2} + \dots + R_j}{\Delta_\psi}.$$

§36. LAGRANGE'S THEOREM, *Continued*

If the function ψ_1 belongs to H as it did before, where H is a subgroup of index k in the group G, and the function φ_1 now belongs to G, we have as corresponding conjugate values of φ_1 and ψ_1 under G:

$$\begin{aligned}&\varphi_1, \varphi_2, \dots, \varphi_{j/k}, \varphi_1, \quad \dots, \varphi_{j/k} \\ &\psi_1, \psi_2, \dots, \psi_{j/k}, \psi_{j/k+1}, \dots, \psi_j;\end{aligned}$$

and we have as equations in the φ_i:

$$\begin{aligned}\varphi_1 + \varphi_2 + \ldots + \varphi_{j/k} + \varphi_1 + \ldots + \varphi_{j/k} &= r_0\\ \psi_1\varphi_1 + \psi_2\varphi_2 + \ldots + \psi_{j/k}\varphi_{j/k} + \psi_{j/k+1}\varphi_1 + \ldots + \psi_j\varphi_{j/k} &= r_1\\ \ldots\ldots\ldots\end{aligned}$$

The φ_i are not all distinct, but this does not interfere with the computation of φ_1 which is to follow, and we have again

$$\varphi_1 = R(\psi_1, c_i)$$

or

$$\varphi_1 = R(\psi_1).$$

This completes the proof of Lagrange's Theorem.

In the last case

$$\psi_1 \neq R(\varphi_1, c_i),$$

for a computation of ψ_1 from equations constructed for the ψ_i leads to determinants vanishing with like columns in the φ_i.

Conversely, if

$$\varphi_1 = R(\psi_1, c_i),$$

then a permutation leaving ψ_1 unaltered leaves unaltered also φ_1, and we infer that the group of φ_1 contains that of ψ_1. If moreover

$$\psi_1 = R(\varphi_1, c_i),$$

then also the group of ψ_1 contains that of φ_1, whence both groups are identical. Thus the Theorem of Lagrange presents a necessary and sufficient condition.[1]

If the function ψ_1 belongs to the subgroup H of G, then G can do no more than permute the conjugate values ψ_i that ψ_1 takes under G. Hence symmetric functions of such values are unaltered by G; yet they may belong to G and may not.

That they may not appears from an example. Conjugate values of a function belonging to identity are:

$\psi_1 = 2x_1 - x_2$	1	↑	↑
$\psi_2 = 2x_2 - x_3$	(123)	G	
$\psi_3 = 2x_3 - x_1$	(132)	↓	S
$\psi_4 = 2x_2 - x_1$	(12)		
$\psi_5 = 2x_3 - x_2$	(13)		
$\psi_6 = 2x_1 - x_3$	(23)		↓

[1] The rest of this paragraph may be omitted on first reading.

These conjugate values are permuted so:

Γ on the ψ_i			S on the x_i
↑	1	$\psi_1\psi_2\psi_3\psi_4\psi_5\psi_6$	1 ↑
	(123)(456)	$\psi_2\psi_3\psi_1\psi_5\psi_6\psi_4$	(123)
	(132)(465)	$\psi_3\psi_1\psi_2\psi_6\psi_4\psi_5$	(132)
	(14)(26)(35)	$\psi_4\psi_6\psi_5\psi_1\psi_3\psi_2$	(12)
	(15)(24)(36)	$\psi_5\psi_4\psi_6\psi_2\psi_1\psi_3$	(13)
↓	(16)(25)(34)	$\psi_6\psi_5\psi_4\psi_3\psi_2\psi_1$	(23) ↓

The symmetric functions

$$\psi_1 + \psi_2 + \psi_3$$

and

$$\psi_4 + \psi_5 + \psi_6$$

of three ψ_i conjugate under G are unaltered by G and conjugate in S. Yet they belong not to G but to S because they are symmetric also in the x_i:[1]

$$\psi_1 + \psi_2 + \psi_3 = \psi_4 + \psi_5 + \psi_6 = x_1 + x_2 + x_3.$$

Another example will occur in §58. Thus we can note:

(52) **A symmetric function of values conjugate under the group G belongs to G or to a group containing G.**[2]

If a function ξ_1 belongs to a group Ξ containing H but no other permutations of G, it behaves under G as a function belonging to H does. For we can arrange the subscripts so that a permutation t_i of G converting

$$\psi_1 \to \psi_i$$

converts also

$$\xi_1 \to \xi_i,$$

and if

$$t_i t_j = t_k,$$

we have

$$\psi_{ij} = \psi_k$$

and also

$$\xi_{ij} = \xi_k.$$

For example, the function

$$\xi_1 = x_1 + x_2 - x_3 - x_4$$

[1] But $\psi_1\psi_2\psi_3$ is not symmetric in the x_i.

[2] If to a group containing G, the corresponding group Γ is imprimitive, by §51.

belonging to

$$\Xi = 1, (12), (34), (12)(34)$$

behaves under

$$G = 1, (23), (24), (34), (234), (243)$$

like the function

$$\psi_1 = x_2 - x_3 - x_4$$

belonging to

$$H = 1, (34).$$

We have:

(1)	(23)	(24)	(34)	(234)	(243)
ψ_1	ψ_2	ψ_3	ψ_1	ψ_2	ψ_3
ψ_2	ψ_1	ψ_2	ψ_3	ψ_3	ψ_1
ψ_3	ψ_3	ψ_1	ψ_2	ψ_1	ψ_2

and this remains correct if we replace ψ by ξ. Permutations outside G replace all ψ_i although not all ξ_i; but they alter any sum of either the ψ_i or the ξ_i. Hence we infer:

(53) **When values taken under G are such that**

(1) the sum of any values[1] conjugate under a group belongs to the group

(2) any value belongs to a group such that

(a) the group contains a subgroup but no other permutations of G

(b) G contains all permutations leaving unaltered or interchanging the values

then the sum of the values belongs to G.

The proposition is true also for other symmetric functions than sums, but is needed mostly for these.[2]

In a sense which we are apt to imply, Lagrange's Theorem fails whenever two conjugate values ψ_i are equal, making the denominator in the expression for φ_1 vanish. We note however that this may not happen in case of the general equation, as will be explained in §70.

[1] Any values of those taken under G.

[2] Compare §§44 and 48.

§37. PLAN OF LAGRANGE

It will be clear now why the solution of a resolvent equation may imply the solution of the general equation

$$f(x) = x^n - c_1 x^{n-1} + \dots \pm c_n = 0$$

for its roots

$$x_1, \dots, x_n.$$

This is so because a root x_1 of the general equation is an n-valued function of the x_i belonging to the subgroup

$$X_1 = S^{n-1}$$

of S^n which acts on all letters x_i other than x_1. Hence x_1 is by Lagrange's Theorem rationally computable from the coefficients c_i of the general equation and any function of the x_i that belongs to X_1 or a subgroup of X_1. But any such function is a root of a resolvent equation.

Since X_1 must contain identity as subgroup, a root of the general equation is certainly computable from the coefficients c_i and a function

$$v = \alpha_1 x_1 + \alpha_2 x_2 + \dots + \alpha_n x_n$$

that belongs to identity and hence is a root of a resolvent equation of degree $n!$.

Suppose that a series of subgroups from the symmetric group S on the n letters x_i down to identity is

$$\begin{array}{ccccccc} S & \leftarrow j \rightarrow & G & \leftarrow k \rightarrow & H & \dots & 1 \\ c_i & & \varphi & & \psi & & v, \end{array}$$

where the indices of the groups are written between them and the functions of the x_i belonging to the groups are written below them. Then φ is a root of a resolvent equation of degree j whose coefficients are rationally computable from the c_i; ψ is a root of a resolvent equation of degree k whose coefficients are rationally computable from φ and the c_i; and so on. We can construct a chain of resolvent equations

$$\varphi^j + R_1(c_i)\varphi^{j-1} + \dots = 0$$
$$\psi^k + R_1'(\varphi, c_i)\psi^{k-1} + \dots = 0$$
$$\dots\dots\dots$$

if need be down to identity and consider these equations solved if their roots can be computed by algebraic operations. This is the plan of Lagrange.

Is our task completed? Alas, here our troubles begin! For the symmetric group on n letters x_i has by proposition (47) no other subgroup than the alternating group of index less than n when n is greater than four; and the alternating group has under such circumstances by proposition (48) no subgroup at all. It follows that among the resolvent equations is one of degree n as the general equation is, and such an equation we cannot solve. Thus Lagrange's tentative plan of solving general equations fails when n is greater than four.

The solution of the general biquadratic equation

$$f(x) = x^4 - c_1x^3 + c_2x^2 - c_3x + c_4 = 0$$

is possible in several ways. One plan of solution[1] is this:

$$\begin{array}{lll} & S & c_i \\ 2\{ & & \\ & A & \sqrt{\Delta} \\ 3\{ & & \\ & V & (x_1 - x_2)(x_3 - x_4) \\ 2\{ & & \\ & H & [\alpha(x_1 - x_2) + \beta(x_3 - x_4)]^2 \\ 2\{ & & \\ & 1 & \alpha(x_1 - x_2) + \beta(x_3 - x_4), \end{array}$$

where the indices of the groups are to their left and the functions belonging to the groups on their right and where[2]

$$H = 1, (12)(34).$$

Since no resolvent equation is of degree more than three, the solution of the general biquadratic equation is reduced to the solution of quadratic and cubic resolvents.

§38. LAGRANGE'S SOLVENT

The solution of the general cubic equation

$$f(x) = x^3 + px - q = 0$$

in chapter two is based on the plan:

$$\begin{array}{lll} & S & p, q \\ 2\{ & & \\ & A & \sqrt{\Delta} \\ 3\{ & & \\ & 1 & \varphi = x_1 + \omega x_2 + \omega^2 x_3. \end{array}$$

[1] Another plan will be found in §46.

[2] V was given in §31.

It is not possible to avoid a resolvent equation of degree three, and yet this resolvent solves the cubic equation which is of the same degree but not solvable without it. By what virtue then can this resolvent render such a service? Evidently because it is binomial making a solution possible by extraction of roots; for it is

$$\varphi^3 = R(\sqrt{\Delta}, q, \omega),$$

the ω coming from the irrational coefficients of φ.

A sudden thought flashes through our minds. If we cannot avoid a resolvent equation of the same degree as the general equation, just when is the resolvent binomial?

If the resolvent equation

$$\psi^j - R(\varphi, c_i) = 0$$

for the values ψ_i conjugate under G of φ is binomial, then

$$\psi_0 = \sqrt[j]{R},\ \psi_1 = \epsilon\sqrt[j]{R},\ \ldots,\ \psi_{j-1} = \epsilon^{j-1}\sqrt[j]{R},$$

where ϵ is the primitive root of unity[1] defined by the formula

$$\epsilon = \cos\frac{2\pi}{j} + i\sin\frac{2\pi}{j}.$$

The conjugate values ψ_i differ by constant factors only and therefore belong to a normal subgroup N of G.

A binomial resolvent either is of prime degree or can be replaced by such resolvents. For assuming that

$$j = p \cdot q$$

where p and q are prime, we can set

$$\psi^q = \chi$$

and

$$\chi^p = R(\varphi, c_i).$$

It follows that the group G of a binomial resolvent[2] either has a normal subgroup of prime index or a series of such:

$$\begin{matrix} G & \overset{p}{\longleftrightarrow} & J & \overset{q}{\longleftrightarrow} & N \\ \varphi & & \chi & & \psi. \end{matrix}$$

Conversely,

(54) **if the group G has a normal subgroup N of prime index, then G is the group of a binomial resolvent for N.**[2]

[1] Cf. §§79 and 84.
[2] Cf. §34, last two lines.

Let the index of N in G be a prime number p, and let ψ_0 be a function of the x_i belonging to N. Under the partitions of

$$G = N + Nt_1 + Nt_2 + \ldots + Nt_{p-1}$$

the conjugate values

$$\psi_0, \psi_1, \psi_2, \ldots, \psi_{p-1}$$

are interchanged in p different ways, as explained in §33, and by proposition (49) the permutations between these conjugate values form a group. This group is circular by proposition (25), and we may set

$$\Gamma = 1, \tau, \tau^2, \ldots, \tau^{p-1}$$
$$\tau = (\psi_0\psi_1 \ldots \psi_{p-1}).$$

We now form the function

$$(\epsilon,\psi)_0 = \psi_0 + \epsilon\psi_1 + \epsilon^2\psi_2 + \ldots + \epsilon^{p-2}\psi_{p-2} + \epsilon^{p-1}\psi_{p-1}$$

also belonging to N, where

$$\epsilon = \cos\frac{2\pi}{p} + i \sin\frac{2\pi}{p}$$

is a primitive root of unity[1] and

$$\epsilon^p = 1.$$

A permutation in the partition Nt_1 interchanges the ψ_i of the function as τ indicates, therefore cyclically, converting $(\epsilon,\psi)_0$ into

$$\begin{aligned}(\epsilon,\psi)_1 &= \psi_1 + \epsilon\psi_2 + \epsilon^2\psi_3 + \ldots + \epsilon^{p-2}\psi_{p-1} + \epsilon^{p-1}\psi_0 \\ &= \epsilon^{-1}(\epsilon,\psi)_0,\end{aligned}$$

while a permutation in the partition Nt_2 interchanges the ψ_i as τ^2 indicates converting $(\epsilon,\psi)_0$ into

$$\begin{aligned}(\epsilon,\psi)_2 &= \psi_2 + \epsilon\psi_3 + \epsilon^2\psi_4 + \ldots + \epsilon^{p-2}\psi_0 + \epsilon^{p-1}\psi_1 \\ &= \epsilon^{-2}(\epsilon,\psi)_0,\end{aligned}$$

and so on. It appears that the permutations of

G on the x_i:	N	Nt_1	Nt_2	....	Nt_{p-1}
like those of Γ on the ψ_i:	1	τ	τ^2	....	τ^{p-1}
convert $(\epsilon,\psi)_0$ into:	$(\epsilon,\psi)_0$	$(\epsilon,\psi)_1$	$(\epsilon,\psi)_2$	....	$(\epsilon,\psi)_{p-1}$
equal to:	$(\epsilon,\psi)_0$	$\epsilon^{-1}(\epsilon,\psi)_0$	$\epsilon^{-2}(\epsilon,\psi)_0$	....	$\epsilon^{-p+1}(\epsilon,\psi)_0$
and to:	$(\epsilon,\psi)_0$	$\epsilon^{p-1}(\epsilon,\psi)_0$	$\epsilon^{p-2}(\epsilon,\psi)_0$	....	$\epsilon(\epsilon,\psi)_0$

[1] Cf. §§79 and 84.

if we multiply by $\epsilon^p = 1$.

As we have

$$(\epsilon,\psi)_i^p = [\epsilon^{p-i}(\epsilon,\psi)_0]^p = (\epsilon,\psi)_0^p,$$

the function $(\epsilon,\psi)_0^p$ is unaltered by any permutation of G and such a permutation alone, so that by Lagrange's Theorem

$$(\epsilon,\psi)_0^p = R(c_i,\varphi,\epsilon).$$

Hence the function $(\epsilon,\psi)_0$ is a root of the binomial resolvent

$$r(\epsilon,\psi) = (\epsilon,\psi)^p - R(c_i,\varphi,\epsilon) = 0,$$

and the group G satisfies the proposition.

The functions $(\epsilon,\psi)_i$ permit the calculation of the ψ_i, as it was illustrated in the solution of the general cubic equation of chapter two by the functions $(\omega,x)_i$, and such a function $(\epsilon,\psi)_i$ is called **Lagrange's solvent**[1] for a normal subgroup N of G.

It follows that

(55) **a resolvent equation for a normal subgroup of prime index is binomial if constructed on Lagrange's solvent for that subgroup.**

It should be noticed that not every function ψ_i belonging to N has its p-th power belong to G as Lagrange's solvent $(\epsilon,\psi)_i$ has.

§39. SPECIAL CASE OF SOLVENTS

For the alternating group of index

$$p = 2$$

in the symmetric group and

$$\epsilon = -1$$

we have by proposition (40)

$$\psi_1 = S_1 + S_2\sqrt{\Delta}$$
$$\psi_2 = S_1 - S_2\sqrt{\Delta},$$

so that Lagrange's solvent is

$$(\epsilon,\psi) = \psi_1 + \epsilon\psi_2 = 2S_2\sqrt{\Delta},$$

in the simplest case just $\sqrt{\Delta}$.

[1] Also called Lagrange's resolvent (function).

Reviewing the solution of the general cubic equation in chapter two, we observe that Lagrange's solvent for the alternating group complies with the rules, but Lagrange's solvent

$$(\omega,x) = x_1 + \omega x_2 + \omega^2 x_3$$

for identity in the alternating group is constructed on the functions

$$x_1 \text{ with group } 1, (23)$$
$$x_2 \text{ with group } 1, (13)$$
$$x_3 \text{ with group } 1, (12)$$

which do not belong to identity. This is possible because the functions x_i belong to groups containing identity but no other permutations of the alternating group

$$A = 1, (123), (132),$$

and because there are no permutations outside the alternating group that leave the functions x_i unaltered.[1]

The significance of the last condition is seen on the function

$$(\omega,\varphi) = \varphi_1 + \omega\varphi_2 + \omega^2\varphi_3$$

with

$$\varphi_1 = x_1x_2 + x_3x_4$$
$$\varphi_2 = x_1x_3 + x_2x_4$$
$$\varphi_3 = x_1x_4 + x_2x_3,$$

which belong to groups[2] containing identity and no other permutations of

$$G = 1, (123), (132)$$

but are unaltered by the permutations of V,[3] which are not in G. The function (ω,φ) behaves under G as if it were Lagrange's solvent for identity in four letters x_i:

1	$\varphi_1 + \omega\varphi_2 + \omega^2\varphi_3 = (\omega,\varphi)$
(132)	$\varphi_2 + \omega\varphi_3 + \omega^2\varphi_1 = \omega^{-1}(\omega,\varphi)$
(123)	$\varphi_3 + \omega\varphi_1 + \omega^2\varphi_2 = \omega^{-2}(\omega,\varphi)$

[1] Compare proposition (53).

[2] These groups were given in §19.

[3] V was given in §31.

Yet this function belongs not to identity but to the group

$$\{1,V\} = V,$$

and its cube not to G but to the group

$$\{G,V\} = A^4.$$

It appears that the functions ψ_i of Lagrange's solvent may conditionally be replaced by functions ξ_i:

(56) **Lagrange's solvents for a normal subgroup N of G can be constructed on conjugate functions belonging to N, or on conjugate functions belonging to groups containing N but no other permutations of G if there are no permutations outside G that leave the conjugate functions unaltered.**

§40. LIMITS OF LAGRANGE'S PLAN

We have learned how to construct binomial resolvents. Does it help us to solve the general equation of degree n? Only when the composition-factors of the symmetric group on n letters x_i are prime.

Whenever the symmetric group in n letters x_i has a series of subgroups each normal and of prime index in the preceding, the series beginning with the symmetric group and ending with identity, then we can solve the general equation of degree n by algebraic operations using primitive roots of unity.[1]

But when these conditions do not hold, it would not seem possible to solve the general equation of degree n by algebraic operations, which beside the rational operations include the extraction of roots.[2]

With scarcely any hope left we therefore acknowledge failure, for we have no such series of normal subgroups with prime indices if n is greater than four, since the symmetric group then has no other normal subgroup than the alternating group and the

[1] We need primitive roots of unity for Lagrange's solvents, but they can be computed, as will be explained in §84.

[2] Compare the statement of our conclusion in §§69 and 76. For the final statement see §82.

alternating group no normal subgroup at all. The composition-series for

$$n > 4$$

is

$$S \leftarrow 2 \rightarrow A \leftarrow \tfrac{n!}{2} \rightarrow 1,$$

where $n!/2$ is not prime.

This seems to put an end to all attempts of solving the general equation of degree higher than four: indeed we shall find[1] that it is not solvable. While special equations of higher degree can be solved, they elude the grip of Lagrange.

That they do, is not a fault. Lagrange's plan of solving equations has been treated as inferior: that it is not. Directed toward the solution of the general equation, it is perfect as such; and concluding our study of it, we salute the genius of the master.

[1] Cf. §76.

CHAPTER VIII

GENERAL EQUATIONS

A. QUADRATIC EQUATION:

$$\boxed{x^2 - c_1x + c_2 = 0}$$

§41. The binomial resolvent is given by the plan

$$\underset{c_i,\Delta}{S} \leftarrow 2 \rightarrow \underset{\sqrt{\Delta},}{A = 1}$$

where the symmetric group is

$$S = 1,\ (12)$$

and the alternating group, normal in the symmetric, shrinks to identity.

Lagrange's solvent for

$$A = 1$$

is[1]

$$\sqrt{\Delta} = x_1 - x_2.$$

Its square, belonging to the symmetric group and hence by Lagrange's Theorem rationally computable from the c_i, is the discriminant

$$\begin{aligned}\Delta &= (x_1 - x_2)^2 \\ &= (x_1 + x_2)^2 - 4x_1x_2 \\ &= c_1{}^2 - 4c_2.\end{aligned}$$

This gives us the binomial resolvent for A.

The roots x_1 and x_2, belonging as functions of the x_i to the group $A = 1$, are by Lagrange's Theorem rationally computable from

$$\sqrt{\Delta} = \sqrt{c_1{}^2 - 4c_2}.$$

We set

$$\begin{aligned}x_1 + x_2 &= c_1 \\ \sqrt{\Delta} = x_1 - x_2 &= \sqrt{c_1{}^2 - 4c_2}\end{aligned}$$

[1] Cf. §39.

and have

$$x_1 = \frac{c_1}{2} + \sqrt{\left(\frac{c_1}{2}\right)^2 - c_2}$$

$$x_2 = \frac{c_1}{2} - \sqrt{\left(\frac{c_1}{2}\right)^2 - c_2}.$$

If we write the quadratic equation in the binomial form:

$$a_0x^2 + 2a_1x + a_2 = 0,$$

and then substitute from

$$c_1 = -\frac{2a_1}{a_0}$$

$$c_2 = \frac{a_2}{a_0},$$

we obtain

$$x_1 = \frac{-a_1 + \sqrt{a_1^2 - a_0a_2}}{a_0}$$

$$x_2 = \frac{-a_1 - \sqrt{a_1^2 - a_0a_2}}{a_0}.$$

B. CUBIC EQUATION:

$$\boxed{x^3 - c_1x^2 + c_2x - c_3 = 0}$$

§42. The solution by binomial resolvents is given with the plan

$$\begin{array}{ccccc} S & \leftarrow 2 \rightarrow & A & \leftarrow .3 \rightarrow & 1 \\ c_i, \Delta & & \sqrt{\Delta}, \varphi_i^3 & & \varphi_i \end{array}$$

where

$$S = 1, (123), (132), (12), (13), (23)$$

and

$$A = 1, (123), (132).$$

The square of Lagrange's solvent for A is the discriminant[1]

$$\Delta = (x_1 - x_2)^2(x_1 - x_3)^2(x_2 - x_3)^2$$

which is symmetric in the x_i and by Lagrange's Theorem rationally computable from the c_i. To do the computation,[2] we notice that the discriminant is homogeneous of degree six in all the x_i and of degree not more than four in any one of them. Hence it is expressible as

$$\Delta = \sum c_1^{\nu_1}c_2^{\nu_2}c_3^{\nu_3},$$

[1] Cf. §39.
[2] Cf. §24, end.

where the weight of each term is

$$W = \nu_1 + 2\nu_2 + 3\nu_3 = 6,$$

while the total degree is

$$D = \nu_1 + \nu_2 + \nu_3 \leqq 4.$$

The combinations of the ν_i satisfying these conditions are

ν_1	ν_2	ν_3
0	0	2
0	3	0
1	1	1
2	2	0
3	0	1.

Consequently

$$\Delta = l_1c_3^2 + l_2c_2^3 + l_3c_1c_2c_3 + l_4c_1^2c_2^2 + l_5c_1^3c_3,$$

where the l_i are numerical coefficients.

We proceed to compute these coefficients with the help of special equations. Thus the equation

(1) $$x^3 - x = 0$$

with

$$x_1 = 0, x_2 = 1, x_3 = -1$$

and

$$c_1 = 0, c_2 = -1, c_3 = 0$$

gives

$$\Delta = -l_2$$

from the equation for Δ, and gives directly

$$\Delta = (0 - 1)^2(0 + 1)^2(1 + 1)^2 = 4,$$

whence

$$l_2 = -4.$$

(2) $$x^3 - 2x^2 + x = 0$$

with

$$x_1 = 0, x_2 = 1, x_3 = 1$$

and

$$c_1 = 2, c_2 = 1, c_3 = 0$$

gives

$$\Delta = l_2 + 4l_4 = 4l_4 - 4 \qquad [l_2 = -4$$

and

$$\Delta = (0 - 1)^2(0 - 1)^2(1 - 1)^2 = 0,$$

whence

$$l_4 = 1.$$

(3) $$x^3 - 3x - 2 = 0$$

with

$$x_1 = 1, x_2 = 1, x_3 = -2$$

gives

$$\Delta = 4l_1 - 27l_2 = 0,$$

whence

$$l_1 = -27.$$

(4) $$x^3 - 3x^2 + 4 = 0$$

with

$$x_1 = 2, x_2 = 2, x_3 = -1$$

gives

$$\Delta = 16l_1 - 108l_5 = 0, \qquad [l_1 = -27$$

whence

$$l_5 = -4.$$

(5) $$x^3 - x^2 - x + 1 = 0$$

with

$$x_1 = 1, x_2 = 1, x_3 = -1$$

gives

$$\Delta = l_1 - l_2 + l_3 + l_4 - l_5 = 0, [l_4 = 1,\ l_5 = -4$$

whence

$$l_3 = 18.$$

Substituting the numerical values of the l_i, we have as binomial resolvent for A the equation

$$\Delta = -27c_3^2 - 4c_2^3 + 18c_1c_2c_3 + c_1^2c_2^2 - 4c_1^3c_3.$$

§43. As Lagrange's solvent for identity we can take by proposition (56) the function

$$(\omega, x) = x_1 + \omega x_2 + \omega^2 x_3 = \varphi_1.$$

Its cube belongs to the alternating group and is by Lagrange's Theorem rationally computable from $\sqrt{\Delta}$: as alternating function it is by proposition (40) of the form

$$\varphi_1^3 = S_1 + S_2\sqrt{\Delta}.$$

To compute it, we have

$$\begin{aligned}\varphi_1^3 &= (x_1 + \omega x_2 + \omega^2 x_3)^3 \\ &= x_1^3 + x_2^3 + x_3^3 + 3\omega(x_1^2x_2 + x_2^2x_3 + x_3^2x_1) \\ &\qquad + 3\omega^2(x_1x_2^2 + x_2x_3^2 + x_3x_1^2) + 6x_1x_2x_3.\end{aligned}$$

Substituting from

$$\omega = \frac{-1 + \sqrt{-3}}{2}$$

$$\omega^2 = \frac{-1 - \sqrt{-3}}{2}$$

and then setting[1]

$$\begin{aligned} s(x_1^3) &= x_1^3 + x_2^3 + x_3^3 \\ s(x_1^2x_2) &= x_1^2x_2 + x_2^2x_3 + x_3^2x_1 + x_1x_2^2 + x_2x_3^2 + x_3x_1^2 \\ s(x_1x_2x_3) &= x_1x_2x_3 \end{aligned}$$

we obtain

$$\varphi_1^3 = s(x_1^3) - \frac{3}{2}s(x_1^2x_2) + \frac{3\sqrt{-3}}{2}(x_1^2x_2 + x_2^2x_3 + x_3^2x_1 - x_1x_2^2 - x_2x_3^2 - x_3x_1^2) + 6s(x_1x_2x_3).$$

Since the third term alone is non-symmetric in the x_i, it must contain $\sqrt{\Delta}$: indeed we find

$$\begin{aligned} \sqrt{\Delta}_{i<k} &= (x_1 - x_2)(x_1 - x_3)(x_2 - x_3) \\ &= x_1^2x_2 + x_2^2x_3 + x_3^2x_1 - x_1x_2^2 - x_2x_3^2 - x_3x_1^2, \end{aligned}$$

so that

$$\varphi_1^3 = s(x_1^3) - \frac{3}{2}s(x_1^2x_2) + 6s(x_1x_2x_3) + \frac{3\sqrt{-3}}{2}\sqrt{\Delta}_{i<k}.$$

Computing now by the method used in §42:

$$\begin{aligned} s(x_1^3) &= 3c_3 - 3c_1c_2 + c_1^3 \\ s(x_1^2x_2) &= -3c_3 + c_1c_2 \\ s(x_1x_2x_3) &= c_3, \end{aligned}$$

or computing directly:

$$\begin{aligned} \varphi_1^3 &= x_1^3 + x_2^3 + x_3^3 + 6x_1x_2x_3 - \frac{3}{2}(x_1^2x_2 + x_2^2x_3 + x_3^2x_1 \\ &\qquad + x_1x_2^2 + x_2x_3^2 + x_3x_1^2) + \frac{3\sqrt{-3\Delta}}{2} \\ &= (x_1 + x_2 + x_3)^3 - \frac{9}{2}(x_1^2x_2 + x_2^2x_3 + x_3^2x_1 + x_1x_2^2 \\ &\qquad + x_2x_3^2 + x_3x_1^2) + \frac{3\sqrt{-3\Delta}}{2} \\ &= (x_1 + x_2 + x_3)^3 - \frac{9}{2}(x_1x_2 + x_2x_3 + x_3x_1)(x_1 + x_2 + x_3) \\ &\qquad + \frac{27}{2}x_1x_2x_3 + \frac{3\sqrt{-3\Delta}}{2}, \end{aligned}$$

[1] Cf. §22.

we have as binomial resolvent for the function

$$\varphi_1 = x_1 + \omega x_2 + \omega^2 x_3$$

belonging to identity the equation

$$\varphi_1^3 = \frac{27}{2}c_3 - \frac{9}{2}c_1c_2 + c_1^3 + \frac{3\sqrt{-3\Delta}}{2}$$

$$= \frac{1}{2}(27\, c_3 - 9c_1c_2 + 2c_1^3 + 3\sqrt{-3\Delta}).$$
$$i<k$$

Applying the transposition (23), we find as binomial resolvent for the function

$$\varphi_2 = x_1 + \omega^2 x_2 + \omega x_3$$

the equation

$$\varphi_2^3 = \frac{1}{2}(27c_3 - 9c_1c_2 + 2c_1^3 - 3\sqrt{-3\Delta}),$$
$$i<k$$

since a transposition changes the sign of $\sqrt{\Delta}$.

The conjugate values of φ_1 under A are

$$\varphi_3 = \omega^2\varphi_1$$
$$\varphi_5 = \omega\varphi_1,$$

the conjugate values of φ_2 under A are

$$\varphi_4 = \omega\varphi_2$$
$$\varphi_6 = \omega^2\varphi_2;$$

and these conjugate values are the remaining roots of the resolvent equations.

§44. We have found the value of every function φ_i belonging to identity, and all what remains for us to do is to compute the value of every function x_i: in terms of any φ_i if we please, by theorem and method of Lagrange in chapter seven, or more conveniently from the sum of two φ_i. For the function φ_1 takes under the group

$$X_1 = 1\ (23)$$

of x_1 the conjugate value φ_2, and the function

$$\varphi_1 + \varphi_2$$

lends itself by proposition (53) to a rational computation of x_1. Recalling that

$$\omega + \omega^2 = -1,$$

we have

$$\begin{array}{r} x_1 + \omega x_2 + \omega^2 x_3 = \varphi_1 \\ x_1 + \omega^2 x_2 + \omega x_3 = \varphi_2 \\ \hline 2\,x_1 - x_2 - x_3 = \varphi_1 + \varphi_2 \\ x_1 + x_2 + x_3 = c_1 \\ \hline 3\,x_1 = c_1 + \varphi_1 + \varphi_2. \end{array}$$

Similarly we may compute any x_i in terms of $\varphi_i + \varphi_k$ properly chosen. Or we may find the x_i adding the same equations as they stand and multiplied by

$$\omega^2, \omega, 1$$

and then by

$$\omega, \omega^2, 1$$

respectively. Recalling that

$$1 + \omega + \omega^2 = 0,$$

we thus have

$$\begin{array}{l} x_1 + \omega x_2 \ + \omega^2 x_3 = \varphi_1 \\ x_1 + \omega^2 x_2 + \omega x_3 \ = \varphi_2 \\ x_1 + \quad x_2 + \quad x_3 = c_1 \\ \hline 3x_1 = \quad c_1 + \quad \varphi_1 + \quad \varphi_2 \\ 3x_2 = \quad c_1 + \omega^2\varphi_1 + \omega\varphi_2 \\ 3x_3 = \quad c_1 + \omega\varphi_1 \ + \omega^2\varphi_2, \end{array}$$

where we may substitute the proper φ_i to rid the result of ω.

§45. If the cubic equation is written in the binomial form:

$$a_0x^3 + 3a_1x^2 + 3a_2x + a_3 = 0,$$

we set

$$c_1 = -\frac{3a_1}{a_0},\ c_2 = \frac{3a_2}{a_0},\ c_3 = -\frac{a_3}{a_0}$$

and obtain

$$\Delta = -\frac{27}{a_0{}^4}(a_0{}^2a_3{}^2 + 4a_0a_2{}^3 - 6a_0a_1a_2a_3 - 3a_1{}^2a_2{}^2 + 4a_1{}^3a_3)$$

$$= -\frac{27}{a_0{}^4}\cdot\frac{(a_0{}^2a_3 - 3a_0a_1a_2 + 2a_1{}^3)^2 - 4(a_1{}^2 - a_0a_2)^3}{a_0{}^2}$$

$$\varphi_1{}^3 = -\frac{27}{2a_0{}^3}(a_0{}^2a_3 - 3a_0a_1a_2 + 2a_1{}^3) + \frac{3\sqrt{-3\Delta}}{2}.$$

Introducing the notation

$$G_2 = a_0a_2 - a_1^2 = \begin{vmatrix} a_0 & a_1 \\ a_1 & a_2 \end{vmatrix}$$

$$G_3 = a_0^2a_3 - 3a_0a_1a_2 + 2a_1^3$$

for functions whose virtues are revealed in the theory of invariants, we finally have

$$\Delta = -\frac{27}{a_0^6}(G_3^2 + 4G_2^3)$$

$$\varphi_1^3 = -\frac{27}{2a_0^3}(G_3 + \sqrt{G_3^2 + 4G_2^3})$$

$$\varphi_2^3 = -\frac{27}{2a_0^3}(G_3 - \sqrt{G_3^2 + 4G_2^3}),$$

and as solution of the cubic equation

$$x_1 = -\frac{a_1}{a_0} - \frac{1}{a_0\sqrt[3]{2}}(G_3 + \sqrt{G_3^2 + 4G_2^3}) - \frac{1}{a_0\sqrt[3]{2}}(G_3 - \sqrt{G_3^2 + 4G_2^3})$$

$$x_2 = -\frac{a_1}{a_0} - \frac{\omega^2}{a_0\sqrt[3]{2}}(G_3 + \sqrt{G_3^2 + 4G_2^3}) - \frac{\omega}{a_0\sqrt[3]{2}}(G_3 - \sqrt{G_3^2 + 4G_2^3})$$

$$x_3 = -\frac{a_1}{a_0} - \frac{\omega}{a_0\sqrt[3]{2}}(G_3 + \sqrt{G_3^2 + 4G_2^3}) - \frac{\omega^2}{a_0\sqrt[3]{2}}(G_3 - \sqrt{G_3^2 + 4G_2^3}).$$

Example:

$$x^3 - 7x^2 + 14x - 8 = 0,$$

$$a_0 = 1,\ a_1 = -\frac{7}{3},\ a_2 = \frac{14}{3},\ a_3 = -8.$$

$$G_2 = a_0a_2 - a_1^2 = -\frac{7}{9}$$

$$G_3 = a_0^2a_3 - 3a_0a_1a_2 + 2a_1^3 = -\frac{20}{27}$$

$$\varphi_1^3 = -\frac{27}{2}(G_3 + \sqrt{G_3^2 + 4G_2^3}) = 10 - 9\sqrt{-3}$$

$$\varphi_1 = -2 - \sqrt{-3}$$

$$\varphi_2 = -2 + \sqrt{-3}$$

$$x_1 + \omega x_2 + \omega^2 x_3 = -2 - \sqrt{-3}$$

$$x_1 + \omega^2 x_2 + \omega x_3 = -2 + \sqrt{-3}$$

$$x_1 + x_2 + x_3 = 7$$

$$x_1 = 1$$

$$x_2 = 2 \qquad [\omega + \omega^2 = -1;\ \omega - \omega^2 = \sqrt{-3}$$

$$x_3 = 4.$$

C. BIQUADRATIC EQUATION:

$$\boxed{x^4 - c_1x^3 + c_2x^2 - c_3x + c_4 = 0}$$

§46. The biquadratic equation written in the binomial form is

$$a_0x^4 + 4a_1x^3 + 6a_2x^2 + 4a_3x + a_4 = 0,$$

where

$$c_1 = -\frac{4a_1}{a_0},\ c_2 = \frac{6a_2}{a_0},\ c_3 = -\frac{4a_3}{a_0},\ c_4 = \frac{a_4}{a_0}.$$

A convenient solution[1] is given by the plan

$$\begin{array}{ccccc} S & \leftarrow 3 \rightarrow & G & \leftarrow 2 \rightarrow & N \\ c_i & & y_1,\ \xi_i^2 & & \xi_i, \end{array}$$

where S is the symmetric group on four letters x_i, as given in §21, while

$$G = 1,\ (12),\ (34),\ (12)(34),\ (13)(24),\ (14)(23),\ (1324),\ (1423)$$

and

$$N = 1,\ (12), (34), (12)(34).$$

A kind fate thus spared us the search for the discriminant.

Since the group G is not normal in S, the cubic resolvent for y_1 is not binomial and has to be solved by one square and one cube root, as shown under B.

Let[2]

$$y_1 = x_1x_2 + x_3x_4.$$

It is a root of the resolvent equation

$$\begin{aligned}(y - y_1)&(y - y_2)(y - y_3) \\ &= y^3 - (y_1 + y_2 + y_3)y^2 + (y_1y_2 + y_2y_3 + y_3y_1)y - y_1y_2y_3 \\ &= A_0y^3 + 3A_1y^2 + 3A_2y + A_3 = 0,\end{aligned}$$

whose other roots are the conjugate values

$$\begin{aligned} y_2 &= x_1x_3 + x_2x_4 \\ y_3 &= x_1x_4 + x_2x_3 \end{aligned}$$

of y_1 under S. The coefficients of this equation are symmetric in the y_i. As a permutation on the x_i can do no more than interchange the y_i, the coefficients are symmetric also in the x_i and hence rationally computable from the c_i.

[1] Cf. §37.
[2] Cf. §§19 and 39.

Noting that

$$A_0 = 1,$$

we compute:

$$-3A_1 = y_1 + y_2 + y_3$$
$$= x_1x_2 + x_3x_4 + x_1x_3 + x_2x_4 + x_1x_4 + x_2x_3 = c_2 = \frac{6a_2}{a_0}.$$

Further:

$$3A_2 = y_1y_2 + y_2y_3 + y_3y_1$$
$$= (x_1x_2 + x_3x_4)(x_1x_3 + x_2x_4) + \ldots = S(x_i),$$

where any term of $S(x_i)$ is of degree four in all the x_i and of degree not more than two in any one x_i. Hence[1]

$$S(x_i) = \sum c_1^{\nu_1}c_2^{\nu_2}c_3^{\nu_3}c_4^{\nu_4},$$

with the condition that

$$W = \nu_1 + 2\nu_2 + 3\nu_3 + 4\nu_4 = 4$$
$$D = \nu_1 + \nu_2 + \nu_3 + \nu_4 \leqq 2.$$

The possible combinations of the ν_i are:

ν_1	ν_2	ν_3	ν_4
0	0	0	1
1	0	1	0
0	2	0	0,

and we have:

$$3A_2 = l_1c_4 + l_2c_1c_3 + l_3c_2^2.$$

To compute the l_i, we use special values of the x_i and the corresponding values of the c_i and the y_i, as we did in §42:

x_1	x_2	x_3	x_4	c_4	c_1c_3	c_2^2	y_1	y_2	y_3	$3A_2$
1	1	0	0	0	0	1	1	0	0	0
1	1	−1	−1	1	0		2	−2	−2	−4
1	1	−1	0	0	−1		1	−1	−1	−1

The first set of values gives

$$3A_2 = l_1c_4 + l_2c_1c_3 + l_3c_2^2 = l_3$$

[1] Cf. §24, end.

and

$$3A_2 = y_1y_2 + y_2y_3 + y_3y_1 = 0,$$

whence

$$l_3 = 0.$$

The second:

$$3A_2 = l_1c_4 + l_2c_1c_3 = l_1 \qquad [l_3 = 0$$

and

$$3A_2 = y_1y_2 + y_2y_3 + y_3y_1 = -4,$$

whence

$$l_1 = -4.$$

The third:

$$3A_2 = l_1c_4 + l_2c_1c_3 = -\, l_2$$

and

$$3A_2 = y_1y_2 + y_2y_3 + y_3y_1 = -1,$$

whence

$$l_2 = 1.$$

Thus we have

$$3A_2 = -4c_4 + c_1c_3 = \frac{-4a_0a_4 + 16a_1a_3}{a_0^2}.$$

Likewise,

$$-A_3 = y_1y_2y_3 = (x_1x_2 + x_3x_4)(x_1x_3 + x_2x_4)(x_1x_4 + x_2x_3)$$
$$= \sum c_1^{\nu_1}c_2^{\nu_2}c_3^{\nu_3}c_4^{\nu_4},$$

with the condition that

$$W = \nu_1 + 2\nu_2 + 3\nu_3 + 4\nu_4 = 6$$
$$D = \nu_1 + \nu_2 + \nu_3 + \nu_4 \leqq 3.$$

The possible combinations of the ν_i are:

ν_1	ν_2	ν_3	ν_4
0	1	0	1
0	0	2	0
0	3	0	0
1	1	1	0
2	0	0	1,

and we have

$$-A_3 = l_1c_2c_4 + l_2c_3^2 + l_3c_2^3 + l_4c_1c_2c_3 + l_5c_1^2c_4.$$

Special values of the x_i give:

x_1	x_2	x_3	x_4	c_2c_4	c_3^2	c_2^3	$c_1c_2c_3$	$c_1^2c_4$	y_1	y_2	y_3	$-A_3$	Result
1	1	0	0	0	0	1	0	0	1	0		0	$l_3 = 0$
1	1	−1	−1	−2	0		0	0	2	−2	−2	8	$l_1 = -4$
1	1	−2	0	0	4		0	0	1	−2	−2	4	$l_2 = 1$
1	1	−1	0	0	1		1	0	1	−1	−1	1	$l_4 = 0$
1	−1	−1	−1	0	4			−4	0			0	$l_5 = 1$

Thus we have

$$-A_3 = -4c_2c_4 + c_3^2 + c_1^2c_4 = \frac{-24a_0a_2a_4 + 16a_0a_3^2 + 16a_1^2a_4}{a_0^3}.$$

§47. We now solve the resolvent equation

$$A_0y^3 + 3A_1y^2 + 3A_2y + A_3$$
$$= y^3 - \frac{6a_2}{a_0}y^2 - \frac{4a_0a_4 - 16a_1a_3}{a_0^2}y + \frac{24a_0a_2a_4 - 16a_0a_3^2 - 16a_1^2a_4}{a_0^3} = 0$$

as explained under B:

$$G_2 = A_0A_2 - A_1^2$$
$$G_3 = A_0^2A_3 - 3A_0A_1A_2 + 2A_1^3,$$

where

$$A_0 = 1, \quad A_1 = \frac{-2a_2}{a_0}$$
$$A_2 = \frac{-4a_0a_4 + 16a_1a_3}{3a_0^2}$$
$$A_3 = \frac{24a_0a_2a_4 - 16a_0a_3^2 - 16a_1^2a_4}{a_0^3}.$$

Substituting for the A_i, we obtain

$$G_2 = -\frac{4}{3a_0^2}(a_0a_4 - 4a_1a_3 + 3a_2^2) = -\frac{4}{3a_0^2}g_2$$
$$G_3 = \frac{16}{a_0^3}(a_0a_2a_4 - a_0a_3^2 - a_1^2a_4 + 2a_1a_2a_3 - a_2^3) = \frac{16}{a_0^3}g_3,$$

where

$$g_2 = a_0a_4 - 4a_1a_3 + 3a_2^2$$
$$g_3 = a_0a_2a_4 - a_0a_3^2 - a_1^2a_4 + 2a_1a_2a_3 - a_2^3 = \begin{vmatrix} a_0 & a_1 & a_2 \\ a_1 & a_2 & a_3 \\ a_2 & a_3 & a_4 \end{vmatrix}$$

The cube of the function[1]

$$\varphi_1 = y_1 + \omega y_2 + \omega^2 y_3$$

is

$$\begin{aligned}\varphi_1^3 &= -\frac{27}{2A_0^3}(G_3 + \sqrt{G_3^2 + 4G_2^3}) \\ &= -\frac{27}{2A_0^3}\left(\frac{16}{a_0^3}g_3 + \sqrt{\frac{256}{a_0^6}g_3^2 - \frac{256}{27a_0^6}g_2^3}\right) \\ &= \frac{-216g_3 - 24\sqrt{81g_3^2 - 3g_2^3}}{a_0^3},\end{aligned}$$

while the cube of the function

$$\varphi_2 = y_1 + \omega^2 y_2 + \omega y_3$$

is

$$\varphi_2^3 = \frac{-216g_3 + 24\sqrt{81g_3^2 - 3g_2^3}}{a_0^3}.$$

Hence

$$\begin{aligned}\varphi_1 &= \frac{2}{a_0}(-27g_3 - 3\sqrt{81g_3^2 - 3g_2^3})^{1/3} \\ \varphi_2 &= \frac{2}{a_0}(-27g_3 + 3\sqrt{81g_3^2 - 3g_2^3})^{1/3},\end{aligned}$$

and we have as solution of the resolvent equation

$$\begin{aligned}y_1 &= \frac{2a_2}{a_0} + \frac{1}{3}(\varphi_1 + \varphi_2) \\ y_2 &= \frac{2a_2}{a_0} + \frac{1}{3}(\omega^2\varphi_1 + \omega\varphi_2) \\ y_3 &= \frac{2a_2}{a_0} + \frac{1}{3}(\omega\varphi_1 + \omega^2\varphi_2).\end{aligned}$$

The discriminant of the resolvent equation is

$$\begin{aligned}\Delta_y &= (y_1 - y_2)^2(y_1 - y_3)^2(y_2 - y_3)^2 \\ &= (x_1x_2 + x_3x_4 - x_1x_3 - x_2x_4)^2(x_1x_2 + x_3x_4 - x_1x_4 - x_2x_3)^2 \\ &\qquad \cdot (x_1x_3 + x_2x_4 - x_1x_4 - x_2x_3)^2 \\ &= (x_1 - x_4)^2(x_2 - x_3)^2(x_1 - x_3)^2(x_2 - x_4)^2(x_1 - x_2)^2 \\ &\qquad \cdot (x_3 - x_4)^2;\end{aligned}$$

[1] Cf. §39.

it is identical with the discriminant Δ_x of the biquadratic equation, which we thus find incidentally, and by §45 we compute:

$$\begin{aligned}\Delta &= -\frac{27}{A_0^{\,6}}(G_3^{\,2} + 4G_2^{\,3}) \\ &= -27\left(\frac{256g_3^{\,2}}{a_0^{\,6}} - \frac{256g_2^{\,3}}{27a_0^{\,6}}\right) \\ &= \frac{256}{a_0^{\,6}}(g_2^{\,3} - 27g_3^{\,2}).\end{aligned}$$

This permits to set

$$\begin{aligned}\varphi_1^{\,3} &= -\frac{216g_3}{a_0^{\,3}} - \frac{3}{2}\sqrt{-3\Delta} \\ \varphi_2^{\,3} &= -\frac{216g_3}{a_0^{\,3}} + \frac{3}{2}\sqrt{-3\Delta}.\end{aligned}$$

The resolvent equation may be brought into the form

$$\left(y - \frac{2a_2}{a_0}\right)^3 - \frac{4g_2}{a_0^{\,2}}\left(y - \frac{2a_2}{a_0}\right) + \frac{16g_3}{a_0^{\,3}} = 0,$$

which is the same as

$$z^3 + 3G_2z + G_3 = 0$$

with

$$z = y - \frac{2a_2}{a_0}.$$

Substituting

$$\frac{2z}{a_0} = y - \frac{2a_2}{a_0},$$

we reduce the resolvent equation to

$$z^3 - g_2z + 2g_3 = 0,$$

and substituting

$$-\frac{4z}{a_0} = y - \frac{2a_2}{a_0},$$

we reduce it to

$$4z^3 - g_2z - g_3 = 0,$$

historic resolvents of the biquadratic equation.

§48. As Lagrange's solvent for the normal subgroup

$$N = 1, (12), (34), (12)(34)$$

of G we take the function

$$(\epsilon, \psi) = \psi_1 + \epsilon\psi_2 = \xi_1,$$

where

$$\epsilon = -1$$
$$\psi_1 = x_1 + x_2.$$

Belonging to N, ψ_1 takes under G the conjugate value

$$\psi_2 = x_3 + x_4,$$

so that

$$\xi_1 = x_1 + x_2 - x_3 - x_4.$$

The square of ξ_1 belongs to G and is by Lagrange's Theorem rationally computable from y_1:

$$\begin{aligned}\xi_1^2 &= (x_1 + x_2 - x_3 - x_4)^2 \\ &= (x_1 + x_2 + x_3 + x_4)^2 - 4(x_1x_3 + x_1x_4 + x_2x_3 + x_2x_4) \\ &= (x_1 + x_2 + x_3 + x_4)^2 - 4(x_1x_2 + x_1x_3 + x_1x_4 + x_2x_3 \\ &\qquad + x_2x_4 + x_3x_4) + 4(x_1x_2 + x_3x_4) \\ &= \left(-\frac{4a_1}{a_0}\right)^2 - 4\frac{6a_2}{a_0} + 4y_1 \\ &= \frac{4}{a_0^2}(4a_1^2 - 6a_0a_2 + a_0^2y_1).\end{aligned}$$

Solving this binomial resolvent for ξ_1, we obtain

$$\xi_1 = x_1 + x_2 - x_3 - x_4 = \pm\frac{2}{a_0}\sqrt{4a_1^2 - 6a_0a_2 + a_0^2y_1}.$$

Although the group N of ξ_1 is not a subgroup of the group

$$X_1 = 1, (23), (24), (34), (234), (243)$$

of x_1, we need not go any further since it contains such a subgroup

$$H = 1, (34)$$

and satisfies both conditions set forth in proposition (53). Hence the sum of the conjugate values that ξ_1 takes under X_1 lends itself to a rational computation of x_1.

The equation for ξ_1 is converted into

$$\xi_2 = x_1 - x_2 + x_3 - x_4 = \pm\frac{2}{a_0}\sqrt{4a_1^2 - 6a_0a_2 + a_0^2y_2}$$

by the permutation (23) of X_1 and into

$$\xi_3 = x_1 - x_2 - x_3 + x_4 = \pm\frac{2}{a_0}\sqrt{4a_1^2 - 6a_0a_2 + a_0^2y_3}$$

by the permutation (24) of X_1. The other three values of ξ_i under the symmetric group on the x_i are the negatives of the values obtained, and from the sum of three ξ_i properly chosen every x_i can be rationally computed.

If we take

$$\begin{aligned} x_1 + x_2 - x_3 - x_4 &= \xi_1 \\ x_1 - x_2 + x_3 - x_4 &= \xi_2 \\ x_1 - x_2 - x_3 + x_4 &= \xi_3 \\ x_1 + x_2 + x_3 + x_4 &= -\frac{4a_1}{a_0}, \end{aligned}$$

we have as solution of the biquadratic equation

$$\begin{aligned} x_1 &= -\frac{a_1}{a_0} + \frac{1}{4}\ (\ \xi_1 + \xi_2 + \xi_3) \\ x_2 &= -\frac{a_1}{a_0} + \frac{1}{4}\ (\ \xi_1 - \xi_2 - \xi_3) \\ x_3 &= -\frac{a_1}{a_0} + \frac{1}{4}(-\xi_1 + \xi_2 - \xi_3) \\ x_4 &= -\frac{a_1}{a_0} + \frac{1}{4}(-\xi_1 - \xi_2 + \xi_3), \end{aligned}$$

where we may replace any negative ξ_i by a positive one if we choose to express an x_i in terms of a sum.

To determine what sign to select for the value of a ξ_i, we notice that any transposition between the x_i leaves one of the three ξ_i unaltered and interchanges, with possible change of sign, the other two. Hence the product $\xi_1\xi_2\xi_3$ remains unaltered under the symmetric group on the x_i and is rationally expressible in terms of the a_i. A computation gives

$$\xi_1\xi_2\xi_3 = -\frac{32}{a_0^3}(a_0^2a_3 - 3a_0a_1a_2 + 2a_1^3),$$

and this is the restriction imposed upon our choice of sign for ξ_i. If in the last calculation we obtain a root x_i that does not satisfy the biquadratic, we have only to change the sign of any one ξ_i to correct our mistake—if we did not blunder before we came to the ξ_i.

Example:

$$x^4 - 2x^3 + 3x^2 + 2x - 4 = 0$$

$$a_0 = 1,\ a_1 = -\tfrac{1}{2},\ a_2 = \tfrac{1}{2},\ a_3 = \tfrac{1}{2},\ a_4 = -4.$$

$$g_2 = a_0a_4 - 4a_1a_3 + 3a_2^2 = -\tfrac{9}{4}$$

$$g_3 = \begin{vmatrix} a_0 & a_1 & a_2 \\ a_1 & a_2 & a_3 \\ a_2 & a_3 & a_4 \end{vmatrix} = -\tfrac{13}{8}$$

$$\varphi_1 = \frac{2}{a_0}\left(-27g_3 - 3\sqrt{81g_3^2 - 3g_2^3}\right)^{1/3} = -3$$

$$\varphi_2 = \frac{2}{a_0}\left(-27g_3 + 3\sqrt{81g_3^2 - 3g_2^3}\right)^{1/3} = 9$$

$$y_1 = \frac{2a_2}{a_0} + \frac{1}{3}(\varphi_1 + \varphi_2) = 3$$

$$y_2 = \frac{2a_2}{a_0} + \frac{1}{3}(\omega^2\varphi_1 + \omega\varphi_2) = 2\sqrt{-3}$$

$$y_3 = \frac{2a_2}{a_0} + \frac{1}{3}(\omega\varphi_1 + \omega^2\varphi_2) = -2\sqrt{-3}$$

$$\xi_1 = \frac{2}{a_0}\sqrt{4a_1^2 - 6a_0a_2 + a_0^2y_1} = -2$$

$$\xi_2 = \frac{2}{a_0}\sqrt{4a_1^2 - 6a_0a_2 + a_0^2y_2}$$

$$= 2\sqrt{-2 + 2\sqrt{-3}} = -2 - 2\sqrt{-3}$$

$$\xi_3 = \frac{2}{a_0}\sqrt{4a_1^2 - 6a_0a_2 + a_0^2y_3}$$

$$= 2\sqrt{-2 - 2\sqrt{-3}} = -2 + 2\sqrt{-3}$$

such that

$$\xi_1\xi_2\xi_3 = -\frac{32}{a_0^3}(a_0^2a_3 - 3a_0a_1a_2 + 2a_1^3) = -32$$

$$\begin{aligned} \xi_1 &= x_1 + x_2 - x_3 - x_4 = -2 \\ \xi_2 &= x_1 - x_2 + x_3 - x_4 = -2 - 2\sqrt{-3} \\ \xi_3 &= x_1 - x_2 - x_3 + x_4 = -2 + 2\sqrt{-3} \\ & x_1 + x_2 + x_3 + x_4 = 2 \end{aligned}$$

$$\begin{aligned} x_1 &= -1 \\ x_2 &= 1 \\ x_3 &= 1 - \sqrt{-3} \\ x_4 &= 1 + \sqrt{-3}. \end{aligned}$$

CHAPTER IX

MORE ABOUT GROUPS

§49. ISOMORPHIC GROUPS

To continue the study of equations, we have to know more about groups and something about domains.

Two groups may be subordinated as group and subgroup, or they may be coördinated by a common property. Coördinated in such a manner that their permutations match under a common law of combination, they are called **isomorphic** and conveniently denoted by G and Γ.

If the permutations of two groups

$$G = 1, s_2, s_3, \ldots, s_r$$
$$\Gamma = 1, \sigma_2, \sigma_3, \ldots, \sigma_r$$

correspond one to one in such a way that

$$\sigma_i \sigma_j = \sigma_k$$

when

$$s_i s_j = s_k,$$

the groups are said to be **simply isomorphic.** Two groups simply isomorphic with a third evidently are simply isomorphic with one another.

Since the product of two transformed permutations is the transformed product of those permutations:

$$t^{-1}s_i t \,.\, t^{-1}s_j t = t^{-1}s_i s_j t,$$

the permutations of every group are in a one to one correspondence with their transforms in the transformed group:

$$G = \ldots, \quad s_i, \quad s_j, \ldots, \quad s_i s_j, \ldots$$
$$t^{-1}Gt = \ldots, t^{-1}s_i t, t^{-1}s_j t, \ldots, t^{-1}s_i s_j t, \ldots$$

Hence

(57) **every group is simply isomorphic with its transform,** and conjugate groups are simply isomorphic with each other.

If the group G is isomorphic with the group Γ in such a way that to the identical permutation in Γ corresponds not one permutation but a set

$$s_1, \ldots, s_n$$

of them in G, this set of permutations is a group since the product

$$s_i\, s_j = s_k$$

of any two such permutations corresponds to

$$1 . 1 = 1$$

in Γ and is in the set by its very correspondence to identity in Γ.

The transform of this group

$$\{s_i\} = s_1, \ldots, s_n$$

by any permutation t of G corresponding to the permutation τ of Γ contains only such permutations of G as correspond to the transform

$$\tau^{-1} . 1 . \tau = 1$$

in Γ and therefore is the group itself:

$$t^{-1}\{s_i\}t = \{s_i\}.$$

It appears that this group is a normal subgroup of G:

$$\{s_i\}_1^n = N.$$

If the permutation t_2 of G corresponds to the permutation τ_2 of Γ, all the permutations Nt_2 do so as they correspond to

$$1 . \tau_2 = \tau_2$$

in Γ, and no other permutations will.[1] It follows that to every permutation in Γ corresponds an equal number of permutations in G. If that number is n, we say that G and Γ are **n to 1 isomorphic,** which we denote by $(n, 1)$-isomorphic. Hence we conclude:

(58) **Whenever two groups G and Γ are $(n, 1)$-isomorphic, we have in correspondence:**

$$\boxed{\begin{array}{l} G = N + Nt_2 + \ldots + Nt_\rho \\ \Gamma = 1, \quad \tau_2, \quad \ldots, \quad \tau_\rho \end{array}},$$

where N is a normal subgroup of G

and ρ is the order of Γ or index of N in G; the order of N is

$$n = \frac{r}{\rho},$$

r being the order of G.

[1] Compare §17.

The permutations on the x_i contained in a partition of G with respect to N are equivalent in their effect upon conjugate functions of the x_i that belong to N or to groups including N as greatest common subgroup.[1] Hence the group G on the x_i is $(n, 1)$-isomorphic with a group Γ of permutations between those functions. It follows from proposition (49) that with N also Γ exists.

An example of such isomorphism occurred in §33 and was tabulated there.

Two groups G and G' may be isomorphic in such a manner that to a normal subgroup of G corresponds a normal subgroup of G'. If they are, the respective groups Γ and Γ' are simply isomorphic.

§50. TRANSITIVE GROUP

The letters x_1 and x_i are said to be connected by a group if there is in the group a permutation that replaces x_1 by x_i, this then implying that there is in the group the inverse permutation replacing x_i by x_1.

If a group connects two letters x_i and x_k with x_1, it connects them with one another; for if it contains a permutation replacing x_i by x_1 and a permutation replacing x_1 by x_k, then it contains also their product which replaces x_i by x_k.

A group connecting x_1 with every one of the other letters is called **transitive.** As it connects all letters with x_1, it connects every letter with every other letter. If it does not do so, it is called **intransitive.** For instance:[2]

$$V = 1, (12)(34), (13)(24), (14)(23)$$

is transitive because its permutations replace 1 by 2 and 3 and 4;

$$W = 1, (12), (34), (12)(34)$$

is intransitive because its permutations replace 1 by 2, but do not replace 1 by either 3 or 4.

(59) **A cyclic group $\{s\}$ is transitive whenever the permutation s is circular,**

as in

$$\{s\} = 1, (1234), (13)(24), (1432)$$

[1] This is true only for a normal subgroup N, by §§32 and 33.

[2] These groups appeared in §31.

with

$$s = (1234)$$

or

$$s = (1432).$$

This is to say that every circular group is transitive.

By the intransitive group W the letters x_1 and x_2 are connected, and so are the letters x_3 and x_4. Thus the four letters x_i of the group are divided by the group into two sets in such a way that each letter is connected with those of its set, but is not connected with those of the other set. Such sets into which the letters of an intransitive group divide are called **intransitive systems**[1] of the group, and the number of letters in a system is its **degree.**

If two transitive groups operate on distinct letters, the group that we obtain by multiplying in all possible ways their permutations is an intransitive group. Thus the intransitive group W is obtained by multiplying the permutations of two transitive groups

$$H = 1, (12)$$

and

$$H' = 1, (34).$$

But we cannot say conversely that every intransitive group is obtained by multiplying the permutations of transitive groups, for

$$G = 1, (12)(34)$$

is not.

While the order of every group of degree n is a divisor of $n!$ by proposition (23), we now prove that

(60) **the order of a transitive group of degree n is a multiple of n.**

Let G be a transitive group of degree n such that its permutation t_i replaces x_1 by x_i, which is to say that

$$t_1 = 1$$

while

$$t_2 \text{ replaces } x_1 \text{ by } x_2$$
$$t_3 \text{ replaces } x_1 \text{ by } x_3$$
$$\cdot \quad \cdot \quad \cdot \quad \cdot \quad \cdot \quad \cdot$$
$$t_n \text{ replaces } x_1 \text{ by } x_n.$$

[1] Also called transitive systems or sets.

Those permutations of G that do not displace x_1—of which there is at least the identical permutation—form a subgroup H of G because the product of two such permutations, not displacing x_1 either, must be in H.

If now t_2 replaces x_1 by x_2, all permutations in Ht_2 evidently do the same, and no other permutations will; all permutations in Ht_3 replace x_1 by x_3, and so on according to the table:

$$\begin{aligned} H &= 1, s_2, \quad \ldots, s_r, \qquad x_1 \to x_1 \\ Ht_2 &= t_2, s_2t_2, \ldots, s_rt_2, \qquad x_1 \to x_2 \\ &\quad \cdots\cdots\cdots \\ Ht_n &= t_n, s_2t_n, \ldots, s_rt_n, \qquad x_1 \to x_n. \end{aligned}$$

This table contains all the permutations of G, since no permutation of G can help leaving x_1 unaltered or replacing it by some x_i. Denoting the order of G by r_g and that of H by r_h, we therefore have

$$r_g = nr_h.$$

If identity alone leaves x_1 unaltered, we have

$$r_h = 1$$

and

$$r_g = n,$$

as for the group V.

A transitive group of degree and order n is called **regular,** because its permutations are all regular. For instance is V a regular group; other examples are the two groups C in §51, while the group W is of degree and order four but not regular.

If all permutations of H leave x_1 unaltered, all permutations of

$$H_i = t_i^{-1}Ht_i$$

leave x_i unaltered since the permutation t_i^{-1} replaces x_i by x_1, the permutations of H do not displace x_1, and the permutation t_i replaces x_1 by x_i. Other permutations of G displace x_i, and it appears that

(61) **the permutations of G which leave any one letter x_i unaltered compose conjugate subgroups of G.**

This is self-evident if we recall the rule of transforms, and we infer that

(62) **any subgroup H_i is transitive in the letters it acts upon if the subgroup H is so.**

A transitive group can be **simply transitive,** as the group V is. It is **doubly transitive** if it connects every pair of letters with every other pair; it can be multiply transitive and, when symmetric of degree n, obviously is n-fold transitive.

Permutations of a doubly transitive group leaving pairs of letters x_i unaltered compose conjugate subgroups, as those do leaving single letters x_i unaltered.

For a group G to be doubly transitive, it is necessary that its subgroup H whose permutations leave one letter x_1 unaltered should be transitive itself in another letter x_i, because the subgroup H must contain a permutation replacing any

$$x_1x_i \rightarrow x_1x_k.$$

And this condition is sufficient since in that subgroup conjugate with H whose permutations leave x_k unaltered there is a permutation replacing

$$x_1x_k \rightarrow x_jx_k,$$

and hence there is in G a permutation replacing

$$x_1x_i \rightarrow x_jx_k.$$

But connecting every pair of letters with x_1x_i, the group G connects every pair of letters with every other pair, and we conclude that

(63) **a group is doubly transitive if its subgroup leaving one letter unaltered is transitive in another.**

The order of the subgroup H is by proposition (60)

$$r_h = (n - 1)r_k,$$

where r_k is the order of a subgroup of H leaving one of the $n - 1$ letters acted upon by H unaltered. Hence the order of a doubly transitive group G is a multiple of n and $n - 1$:

$$r_g = n(n - 1)r_k.$$

§51. IMPRIMITIVE GROUP

Transitive groups may be primitive or imprimitive. If the letters operated on by a transitive group divide into equal sets in such a way that the permutations of the group either interchange the letters within the sets or replace them by the letters of other sets but never break up the sets, then the group is called

imprimitive or non-primitive, and the sets are called **imprimitive systems** of the group. Otherwise the group is called **primitive.**

For instance, the transitive group

$$C = 1, (1234), (13)(24), (1432)$$

on four letters x_i is imprimitive because it has the imprimitive systems

$$x_1, x_3 \mid x_2, x_4;$$

but any alternating group

$$A = 1, (123), (132), \ldots$$

is primitive because it has no such systems.

Those permutations of an imprimitive group G that do not displace the imprimitive systems form a subgroup of G, since the product of any two such permutations does not displace the imprimitive systems either. This subgroup of permutations only interchanging the letters within the imprimitive systems is normal in G by proposition (58), so that we may denote it by N and set

$$G = N + Nt_2 + \ldots + Nt_\rho.$$

For the permutations of this subgroup correspond with identity in the group

$$\Gamma = 1, \tau_2, \ldots, \tau_\rho$$

composed of permutations between the imprimitive systems themselves, as caused by the permutations of G between the x_i, and obviously isomorphic with G. In the example given above we have

$$N = 1, (13)(24).$$

Since the normal subgroup N operates on letters of different imprimitive systems but does not connect them it is intransitive, and we have the proposition:

(64) **The permutations of an imprimitive group not displacing its imprimitive systems form a normal subgroup which is intransitive.**

Conversely,

(65) **if a transitive group has a normal subgroup which is intransitive, then the group is imprimitive.**

Suppose that t is a permutation in the transitive group G connecting two letters in different intransitive systems of the normal subgroup J of G. The transform $t^{-1}Jt$ is obtained by proposition (29) if we operate the permutation t within the cycles of J. But

$$t^{-1}Jt = J$$

and consequently has the same intransitive systems. Hence the permutation t, having replaced in the cycles of J a letter of one intransitive system by a letter of another, must have so replaced every letter interchanging the two systems. This means that the permutations of G either leave the intransitive systems of J unaltered, as the permutations of J do, or interchange them, as the permutation t does, whence it appears that G is imprimitive as our proposition states.

Since N contains all permutations which do not displace the imprimitive systems of G, it follows that J is a subgroup of N if it is not identical with it.

It is clear that the intransitive systems of J or N are also the imprimitive systems of G.

No imprimitive group can be more than simply transitive, for its permutations can replace a pair of letters which is contained in an imprimitive system not by any other pair of letters, but by such pairs alone as belong to some imprimitive system.

If the letters x_i operated on by the permutations of a group divide into k imprimitive systems with m letters each, the letters of any one system can interchange in $m!$ ways, which gives for all systems $(m!)^k$ combinations; and the systems can interchange in $k!$ ways. Hence the greatest number of permutations possible for imprimitive groups with such systems is

$$r = (m!)^k . k!$$

In case of four letters x_i the greatest imprimitive group with the imprimitive systems

$$x_1, x_2 \mid x_3, x_4$$

has the order

$$r = 8$$

and is the group of the function

$$\psi = x_1x_2 + x_3x_4$$

given in §17.

An imprimitive group may be imprimitive in more than one way. For instance, the transitive group
$C = 1, (123456), (135)(246), (14)(25)(36), (153)(264), (165432)$
on six letters x_i has the imprimitive systems

$$x_1, x_3, x_5 \mid x_2, x_4, x_6$$

with the normal and intransitive subgroup

$$N = 1, (135)(246), (153)(264);$$

and also the imprimitive systems

$$x_1, x_4 \mid x_2, x_5 \mid x_3, x_6$$

with the normal and intransitive subgroup

$$N' = 1, (14)(25)(36).$$

§52. QUOTIENT-GROUP

If we multiply two groups H and H', which is to say multiply in all possible ways their permutations, we obtain a definite group

$$G = \{H, H'\}.$$

We call G the **product** of H and H', and both H and H' are contained in G as subgroups.

But the order of G is not in general the product of the orders of H and H'. For instance: if

$$H = 1, (12)$$

and

$$H' = 1, (34),$$

then

$$G = 1, (12), (34), (12)(34);$$

if

$$H = 1, (12)$$

and

$$H' = 1, (12)(34),$$

then

$$G = 1, (12), (12)(34), (34),$$

the order being equal to the product in both cases. But if

$$H = 1, (12)$$

and

$$H' = 1, (13),$$

we have[1]

$$G = 1, (12), (13), (23), (123), (132)$$

with an order greater than the product, while we have

$$G = \{H, H'\} = H$$

if H' is a subgroup of H, the order now being less than the product.

The product of two groups is called **direct** if every permutation of one group is commutative with every permutation of the other and the two groups have no permutation but identity in common. It is readily seen that

(66) **the order of the direct product of two groups is the product of their orders.**

For instance, the product of

$$H = 1, (12)$$

and

$$H' = 1, (34)$$

is direct.

Pursuing the analogy with numbers, we may well inquire whether division of groups has a meaning.

If we arrange the permutations of

$$G = H + Ht_2 + \ . \ . \ . + Ht_r$$

into partitions with respect to H, the permutations

$$1, t_2, \ . \ . \ . \ , t_r$$

of H' do not necessarily form a group. Indeed, it may not even be possible to pick them out so that they do form a group. And again, it may be possible to pick out more than one group H'. Thus for

$$G = \begin{cases} 1, \quad (12), \quad (34), \quad (12)(34) = H \\ (13)(24), (1423), (1324), (14)(23) = Ht_2, \end{cases}$$

we find

$$H' = 1, (13)(24)$$

or

$$H' = 1, (14)(23).$$

All then we can say is that in every partition Ht_i of a group G with respect to its subgroup H there may be contained a permutation q_i such that we can set

$$G = H + Hq_2 + \ . \ . \ . + Hq_\rho$$

[1] Symmetric group by proposition (33).

and that the permutations q_i compose a group. This group we call the **quotient** of G divided by H and denote by Q:

$$Q = 1, q_2, \ldots, q_\rho.$$

The quotient is in general neither certain nor unique, but evidently it is so when G is the direct product of H and H', for then

$$Q = H'.$$

Here the matter ends unless the subgroup is normal. In that case the quotient is not certain either; and it is not any more definite as we see from the example:

$$G = \begin{cases} 1, (12)(34), (13)(24), (14)(23) = N \\ (12), (34), \quad (1324), \quad (1423) \quad = Nt_2, \end{cases}$$

where

$$Q = 1, (12)$$

or

$$Q = 1, (34).$$

But if the quotient of

$$G = N + Nt_2 + \ldots + Nt_\rho$$

divided by N, say

$$Q = 1, q_2, \ldots, q_\rho,$$

exists, it is isomorphic with G in such a manner that its identical permutation corresponds to N in G, and it is therefore simply isomorphic with an always existing[1] and readily constructed group

$$\Gamma = 1, \tau_2, \ldots, \tau_\rho$$

of permutations between conjugate functions belonging to N. This gives a good deal of information about Q since simply isomorphic groups obey the same law of combination.

If we emancipate our notion of groups from its connection with permutations and regard a group as exhibiting on elements, whether permutations or otherwise, a definite law of combination for which alone it stands, we obtain the concept of an **abstract group.**[2]

Abstractly speaking, simply isomorphic groups are identical; and the quotient Q, together with the group Γ, merges into an abstract group called the **abstract quotient** of G divided by N, or

[1] Cf. §49, near end.

[2] Cf. the note at the end of this chapter.

factor-group, and denoted by G/N. We take it as composed of the elements Nt_i; they contain the product of any two elements among them, since by proposition (32)

$$Nt_i \,.\, Nt_j = NNt_it_j = Nt_k,$$

and contain N as the identical element:

$$G/N = N, Nt_2, \ldots, Nt_\rho$$
$$Q = 1, \quad q_2, \quad \ldots, \quad q_\rho$$
$$\Gamma = 1, \quad \tau_2, \quad \ldots, \quad \tau_\rho.$$

With the group Γ, the factor-group G/N always exists, and we can note:

(67) **If a group G has a normal subgroup N, there exists the factor-group G/N abstractly identical with the group Γ of permutations operated by G on functions belonging to N or to groups containing N as greatest common subgroup. The order of G/N is equal to the index of N in G.**

By proposition (25) it follows that G/N is cyclic if the index of N in G is prime, and this obviously applies also to Q and Γ.

§53. SUBGROUPS OF QUOTIENT-GROUP

To every subgroup of the quotient Q that we obtained dividing G by N corresponds isomorphically a subgroup of G containing N, and a similar relation holds true for Γ.

To verify this, let a subgroup of Q be

$$R = 1, q_\alpha, q_\beta, \ldots$$

Then the partitions of G with respect to N which correspond to the permutations of R evidently compose a subgroup

$$H = N + Nt_\alpha + Nt_\beta + \ldots$$

of G containing N. But R is isomorphic with H in such a manner that its identical permutation corresponds to N in H, whence its law of combination is given by the factor-group

$$H/N = N, Nt_\alpha, Nt_\beta, \ldots$$

If R is normal in Q, also H is normal in G, since the transform of Nt_α by a permutation of G corresponds to the transform of q_α by a permutation of Q, and with this transform in R the other is in

H. In this case G is isomorphic with Q in such a manner that the permutations of H in G correspond to the permutations of R in Q:

$$G = H + Ht_a + Ht_b + \dots$$
$$Q = R + Rq_a + Rq_b + \dots,$$

for with[1]

$$Ht_a \,.\, Ht_b = Ht_at_b$$

also

$$Rq_a \,.\, Rq_b = Rq_aq_b,$$

and

$$q_aq_b = q_c$$

corresponds to

$$t_at_b = t_c.$$

It follows that

$$Q/R = G/H,$$

and the index of R in Q equals the index of H in G.

Replacing Q and R by abstract groups, we may write

$$\{G/N\}/\{H/N\} = \{G/N\},$$

which agrees with the rules as to division of elementary algebra, and we have the proposition:

(68) **To every subgroup of G/N corresponds a subgroup of G containing N; if one is normal, the other is also.**

§54. MAXIMUM NORMAL SUBGROUP

A normal subgroup of G contained in no other normal subgroup of G is called a **maximum normal subgroup** of G and may be denoted by $\overline{N}$.

This does not imply that there is in G no normal subgroup of greater or the same order, only that it does not contain a maximum normal subgroup. The group

$G = 1, (12), (34), (12)(34), (13)(24), (14)(23), (1324), (1423),$

for instance, has two maximum normal subgroups:

$$\overline{N} = 1, (12)(34), (13)(24), (14)(23)$$
$$\overline{N}' = 1, (12), (34), (12)(34).$$

To maximum normal subgroups applies the proposition:

[1] Cf. proposition (32).

(69) If $\overline{N}$ and $\overline{N}'$ are maximum normal subgroups of G and D is their greatest common subgroup, then D is a maximum normal subgroup of both $\overline{N}$ and $\overline{N}'$ for which

$$G/\overline{N} = \overline{N}'/D$$

and

$$G/\overline{N}' = \overline{N}/D.$$

To prove this, we notice that $\overline{N}\overline{N}'$ is a normal subgroup of G, since any permutation t of G gives

$$t^{-1}\overline{N}\overline{N}'t = t^{-1}\overline{N}t \,.\, t^{-1}\overline{N}'t = \overline{N}\overline{N}';$$

and since $\overline{N}\overline{N}'$ contains the maximum normal subgroups $\overline{N}$ and $\overline{N}'$, it even is identical with G:

$$G = \overline{N}\overline{N}'.$$

Setting

$$\overline{N}' = D + Dt_2 + \ . \ . \ . + Dt_\rho,$$

we have

$$G = \overline{N}\overline{N}' = \overline{N}(D + Dt_2 + \ . \ . \ . + Dt_\rho)$$

or

$$G = \overline{N} + \overline{N}t_2 + \ . \ . \ . + \overline{N}t_\rho,$$

because D is a subgroup of $\overline{N}$ and therefore

$$\overline{N}D = \overline{N}.$$

The partitions of G with respect to $\overline{N}$ are all distinct as from

$$\overline{N}t_i = \overline{N}t_k$$

would follow

$$\overline{N}t_it_k^{-1} = \overline{N},$$

showing that $t_it_k^{-1}$ is a permutation of $\overline{N}$. Since $t_it_k^{-1}$ also is in $\overline{N}'$ which contains any combination of the t_i, it would have to be in D. But

$$Dt_it_k^{-1} = D$$

gives

$$Dt_i = Dt_k$$

which is untrue.

Now with

$$\overline{N}\,t_i \,.\, \overline{N}t_j = \overline{N}t_it_j = \overline{N}t_k$$

also

$$Dt_i \,.\, Dt_j = Dt_it_j = Dt_k,$$

for D is normal by proposition (39),[1] and it follows that the groups

$$G = \overline{N} + \overline{N}t_2 + \ . \ . \ . + \overline{N}t$$

[1] The proof of proposition (39) given for conjugate subgroups remains true for maximum normal subgroups.

and

$$\overline{N}' = D + Dt_2 + \ldots + Dt_\rho$$

are isomorphic in such a manner that the permutations of D in $\overline{N}'$ correspond with the permutations of $\overline{N}$ in G, whence

$$G/\overline{N} = \overline{N}'/D.$$

Interchanging N and $\overline{N}'$, we prove likewise that

$$G/\overline{N}' = \overline{N}/D.$$

All these factor-groups are simple since by proposition (68) it appears that only when $G/\overline{N}$ is simple has G no normal subgroup containing $\overline{N}$ and is $\overline{N}$ a maximum normal subgroup of G.

Again, since these factor-groups are simple is D a maximum normal subgroup of both $\overline{N}$ and $\overline{N}'$.

§55. CONSTANCY OF COMPOSITION-FACTORS

The proposition we proved leads to an important theorem due to **Jordan and Hölder:**

(70) **If a group has more than one composition-series, their factor-groups are identical except for the sequence.**

Suppose the proposition is true for groups whose order is the product of fewer than n prime numbers, and let G be a group whose order is the product of n prime numbers. If we can prove that the proposition is true for G, it will by induction be true in general, since the constancy of factor-groups is evident for groups of prime order, which are simple.

Let two composition-series of G be

$$G \quad N \quad I \quad \ldots \quad 1$$

and

$$G \quad N' \quad I' \quad \ldots \quad 1,$$

made up of maximum normal subgroups by the definition of a composition-series.

Let the greatest common subgroup of N and N' be D; then we can construct by proposition (69) two other composition-series of G containing D and identical from D on:

$$G \quad N \quad D \quad J \quad \ldots \quad 1$$
$$G \quad N' \quad D \quad J \quad \ldots \quad 1,$$

having the factor-groups

$$G/N \quad N/D \quad D/J \quad \ldots$$
$$G/N' \quad N'/D \quad D/J \quad \ldots$$

But

$$G/N = N'/D$$

and

$$G/N' = N/D$$

by proposition (69), and it appears that the last two composition-series have identical sets of factor-groups with the sequence of the first two factor-groups inverted.

Again, the two composition-series

$$N \quad I \quad \ldots \quad 1$$

and

$$N \quad D \quad \ldots \quad 1$$

as well as the two composition-series

$$N' \quad I' \quad \ldots \quad 1$$

and

$$N' \quad D \quad \ldots \quad 1$$

have identical factor-groups by assumption, since the order of G is the product of n primes and consequently the order of both N and N' the product of fewer than n primes.

It follows that with the two composition-series

$$G \quad N \quad D \quad \ldots \quad 1$$

and

$$G \quad N' \quad D \quad \ldots \quad 1$$

also the two composition-series

$$G \quad N \quad I \quad \ldots \quad 1$$

and

$$G \quad N' \quad I' \quad \ldots \quad 1$$

have identical factor-groups, which proves the theorem.

The existence for the symmetric group of a composition-series with prime composition-factors is a sufficient condition for the solvability of the general equation, as we know. The theorem of Jordan-Hölder relieves us of the necessity to investigate all possible composition-series, since they all have the same composition-factors.

In a sense which is obvious for the symmetric group, and which will become so for other groups in connection with the theory of

Galois, we call a group whose factors of composition are all prime a **soluble group.**[1] Hence we conclude that

(71) **all factor-groups in the composition-series of a soluble group are cyclic;**

this is true by proposition (25), for the order of any such factor-group is prime by proposition (67).

§56. ABELIAN GROUP

Just as the transform of a normal subgroup of the group G by any permutation of G is the same subgroup, so it may happen that the transform of a permutation z of G by any permutation t of G is the same permutation z. Such a permutation is called **normal** in G.

If z' is another normal permutation of G, then also the product zz' is normal in G, for

$$t^{-1}zz't = t^{-1}zt \,.\, t^{-1}z't = zz'.$$

Hence all the normal permutations of G form a subgroup, and this subgroup, normal not only as a whole but in every permutation, is called the **central subgroup**[2] of G and may be denoted by Z. It is needless to say that every subgroup of Z is normal in G. A trivial case of a central subgroup presents itself in identity.

Being normal in G,

(72) **the permutations of the central subgroup of G, and such permutations alone, are commutative with every permutation of G,**

since from

$$t^{-1}zt = z$$

follows that

$$zt = tz,$$

and conversely.

If the permutations of a group are all commutative, we call the group **commutative or Abelian,** in memory of a man who excelled in genius and misfortune.[3] It is clear that subgroups of an Abelian

[1] A soluble group is also called a **metacyclic group,** and a solvable equation a **metacyclic equation.** This has to be remembered well if one is to understand the literature, but we shall use the term in another sense.

[2] Called "Zentrum" in German.

[3] Abel lived 1802–1829. And when he died, for lack of means to live, his government erected him a monument.

group also are Abelian. Denoting an Abelian group by A or $\langle A\rangle$, we observe that

(73) **the central subgroup is Abelian:**

$$\boxed{\{Z\} = \langle A\rangle}.$$

Every subgroup of an Abelian group, indeed every permutation, is normal since

$$t^{-1}st = t^{-1}ts = s$$

for any permutations s and t of the group. Therefore we note that

(74) **an Abelian group is its own central subgroup.**

A simple case of an Abelian group is a cyclic group: the cyclic group

$$\{s\} = s, s^2, \ldots s^r \qquad [s^r = 1$$

is Abelian since

$$s^i s^j = s^j s^i = s^{i+j}.$$

It is quickly verified that

(75) **every cyclic group of order r has a cyclic subgroup, and only one, of every order which is a divisor of r.**

In particular, if r has a prime factor p and

$$kp = r,$$

the cyclic group $\{s\}$ has the cyclic subgroup

$$\{s^p\} = s^p, s^{2p}, \ldots, s^{kp} \qquad [s^{kp} = 1$$

of order k and index p. Hence it follows that

(76) **every cyclic group is soluble.**

In general it is true that

(77) **an Abelian group whose order is divisible by a prime number p contains a permutation of order p.**

For suppose the Abelian group A is generated by the permutations

$$s_1, s_2, s_3, \ldots$$

and let r_1 be the order of s_1. If then r_2 is the order of s_2, the permutation $s_2{}^{r_2}$ obviously appears in $\{s_1\}$ because

$$s_2{}^{r_2} = 1.$$

But it may happen that already a lower power of s_2 appears in $\{s_1\}$, determining what is called the **order of s_2 relative to $\{s_1\}$.**

Let it be ρ_2. The order ρ_2 of s_2 relative to $\{s_1\}$ is a divisor of the **absolute order** r_2 of s_2, for all the permutations

$$s_2^{\rho_2}, s^{2\rho_2}, s^{3\rho_2}, \ldots$$

and these alone,[1] are in $\{s_1\}$; but so is $s_2^{r_2}$, which therefore is one of them, and

$$r_2 = l\rho_2.$$

Similarly, let ρ_3 be the order of s_3 relative to $\{s_1, s_2\}$ while r_3 is the absolute order of s_3, and so on. The order r of A then is

$$r = r_1\rho_2\rho_3 \ldots,$$

and all permutations of A can be expressed in the form

$$s_1^{i_1}s_2^{i_2}s_3^{i_3} \ldots$$

where

$$i_1 = 1, 2, \ldots, r_1$$
$$i_j = 1, 2, \ldots, \rho_j$$
$$j = 2, 3, \ldots$$

If now the order r of A is divisible by a prime number p, it follows that divisible by p is also r_1 or some ρ_i which itself divides r_i. For some r_i we thus have

$$r_i = kp,$$

and s_i^k is a permutation in A of order p since

$$(s_i^k)^p = 1.$$

This enables us to prove that

(78) **an Abelian group of order r contains a subgroup of any order h which is a divisor of r.**

The proposition is true for any order whose factors are all prime, since corresponding to any such factor there is by proposition (77) a permutation in the Abelian group A whose powers compose such a subgroup.

We assume that the proposition is true for any order which is smaller than r; if we can prove that it is true also for r, it is by induction true in general.

Let p be a prime factor of h and hence of r, and let a permutation of order p in A be s. The quotient-group $A/\{s\}$ is Abelian of order[2] r/p, which is smaller than r, and by assumption it

[1] Since otherwise a power of s_2 lower than ρ_2 would be in $\{s_1\}$.

[2] Cf. proposition (67).

contains a subgroup of order h/p. But to this subgroup of $A/\{s\}$ corresponds in A by proposition (68) a subgroup of index

$$\frac{r}{p} \div \frac{h}{p} = \frac{r}{h}$$

and hence of order h, which proves our proposition.

It follows that not only every cyclic group, but

(79) **every Abelian group is soluble.**

If we write the order of A as

$$r = p_1^{\alpha_1} p_2^{\alpha_2} \ . \ . \ . \ p_k^{\alpha_k},$$

where every p_i is prime, it appears from proposition (78) that A has a subgroup H_i of every order $p_i^{\alpha_i}$, and only one. For if it had two, their product would be a subgroup of order $p_i^{\beta_i}$ with $\beta_i > \alpha_i$, and $p_i^{\beta_i}$ would have to divide r which is impossible.

The different subgroups H_i have no permutation but identity in common, since any subgroup H_i contains only permutations whose order is a power of p_i, so as to divide the order $p_i^{\alpha_i}$ of H_i. The direct product of the H_i has the order

$$p_1^{\alpha_1} p_2^{\alpha_2} \ . \ . \ . \ p_k^{\alpha_k} = r$$

by proposition (66), and having the same order as A it is identical with A. Hence we conclude that

(80) **every Abelian group is the direct product of Abelian subgroups whose orders are powers of prime numbers.**

If

$$H_1 = \{s_1\},\ H_2 = \{s_2\},\ . \ . \ .$$

and

$$s = s_1 s_2 \ . \ . \ .,$$

then

$$A = \{s\}.$$

Therefore A is cyclic if the H_i are so.

§57. THEOREM OF CAUCHY

Permutations which are conjugate under a group G make up what we call a **class of permutations** in G. It appears that

(81) **a group is composed of classes which have no common permutations; the permutations of a class have the same order,**

since conjugate permutations are similar by proposition (30).

(82) **Permutations of a group G which are commutative with a given permutation c of G form a subgroup of G whose index equals the number of permutations in the class of c.**

They form a subgroup H because with the permutations s_1 and s_2 contained in H also their product s_1s_2 qualifies for H, since

$$s_1s_2 \,.\, c = s_1cs_2 = c \,.\, s_1s_2.$$

If two permutations t_1 and t_2 which are in G but not in H transform c into the same conjugate permutation, then

$$t_1^{-1}ct_1 = t_2^{-1}ct_2$$

or

$$c \,.\, t_1t_2^{-1} = t_1t_2^{-1} \,.\, c.$$

Therefore the permutation $t_1t_2^{-1}$ is in H, and from

$$Ht_1t_2^{-1} = H$$

or

$$Ht_1 = Ht_2$$

we infer that G has as many partitions with respect to H as there are permutations conjugate to c in G. But this is to say that the index of H in G equals the number of permutations in the class of c.

It follows that the number of permutations in the class of c divides the order of G, since the index of H does. We may note that the subgroup H of G is called the **normalizer** of c in G.

If every permutation of G is commutative with the permutation c, the number of permutations in the class of c is one, which is to say that c is a normal permutation as we should expect it to be by proposition (72). A trivial case presents itself when the permutation c is identity.

Now we can prove the **Theorem of Cauchy**:[1]

(83) **If the order of a group is divisible by a prime number p, the group contains a permutation of order p.**

Suppose the proposition is true for groups whose order is the product of fewer than n prime numbers, as we know it to be true by proposition (25) for groups of order p. We proceed to prove that it then is true also for groups whose order is the product of n prime numbers.

[1] The Theorem of Cauchy was stated without proof by Galois.

Let G be such a group. If it contains a subgroup whose index is prime to p, then the order of the subgroup is divisible by p and the subgroup contains by assumption a permutation of order p.

Otherwise the index of every subgroup is divisible by p. By proposition (82) the number of permutations in any class of G then is divisible by p, too, if it is not one. Also the sum of such numbers is divisible by p, because it is by proposition (81) equal to the order of G. Since the number for the class identity is one, we conclude that there must be more classes with just one permutation and altogether a multiple of p such classes.

The permutations of these classes compose the central subgroup of G. This subgroup is Abelian by proposition (73) and, having an order equal to a multiple of p, contains by proposition (77) a permutation of order p.

Thus one subgroup of G must contain a permutation of order p, which proves the theorem.

§58. METACYCLIC GROUP

While we traced the composition-series of the symmetric group from that group down to identity and found that no symmetric group of degree more than four is soluble, we now reverse the procedure: tracing the composition-series from identity up to larger groups, we search for the largest soluble group within our reach.

But in doing so, we restrict our purpose in two ways:

(1) we confine our work to transitive groups

(2) we confine our work to groups of prime degree.

The soluble group next to identity then is the cyclic group

$$C = \{s\}$$

of order p with

$$s = \begin{pmatrix} 1\,2 \ldots p \\ 2\,3 \ldots 1 \end{pmatrix} = (12 \ldots p).$$

In an other notation, with the modulus p understood, we write

$$s = \begin{pmatrix} z \\ z+1 \end{pmatrix};$$

which means that every subscript

$$z = 1, 2, \ldots, p$$

is replaced by the subscript $z + 1$ and p by 1 since

$$p \equiv 0 (\text{mod } p).$$

In this notation we obviously have

$$s^2 = \binom{z}{z+2}, \quad s^3 = \binom{z}{z+3}, \quad \ldots$$

so that all permutations in C are of the form

$$s^\lambda = \binom{z}{z+\lambda}$$

with

$$\lambda = 1, 2, \ldots, p.$$

Any transitive group contains the cyclic as a subgroup, by the Theorem of Cauchy.[1] And the cyclic group immediately precedes identity in the composition-series of a transitive group, if the transitive group is soluble. For

(84) **a normal subgroup of a transitive group of prime degree is transitive, unless it is identity;**

otherwise a transitive group of prime degree p would have to be imprimitive by proposition (65), but p letters cannot divide into equal sets. It follows that every transitive group has a transitive composition-series, by which we mean a composition-series formed by transitive groups alone.

Hence we search for the largest group containing the cyclic as a normal subgroup. That group beyond the cyclic we call the **metacyclic group,**[2] using the term in its original and literal meaning, and we denote the group by M. The composition-series of the metacyclic group is

$$M \quad \ldots \quad C \quad 1.$$

Suppose that

$$t = \begin{pmatrix} 12 \ldots p \\ ab \ldots k \end{pmatrix}$$

is any permutation of M which is not in C. To represent it in the other notation as

$$t = \binom{z}{\varphi(z)},$$

we need a function $\varphi(z)$ such that

$$\varphi(1) = a, \quad \varphi(2) = b, \quad \ldots, \quad \varphi(p) = k.$$

[1] The order of a transitive group is a multiple of p, by proposition (60).

[2] The metacyclic group is denoted as **linear group** or **congruence group** when metacyclic is used in the sense of soluble, and these terms are applied to include the subgroups of the metacyclic group. This has to be remembered well if one is to understand the literature.

Such a function is given by the interpolation formula of Lagrange:[1]

$$\varphi(z) = \frac{af(z)}{(z-1)f'(1)} + \frac{bf(z)}{(z-2)f'(2)} + \cdots + \frac{kf(z)}{(z-p)f'(p)},$$

where

$$f(z) = (z-1)(z-2)\ldots(z-p)$$

and $f'(z)$ is the derivative of $f(z)$.

Since C is normal in M, the transform of a permutation s of C by a permutation t of M is contained in C and hence a power of s:

$$t^{-1}st = s^{\mu};$$

here

$$\mu \neq 1$$

because s is not identity. But we have:

$$t^{-1} = \begin{pmatrix} \varphi(z) \\ z \end{pmatrix}$$

$$t^{-1}s = \begin{pmatrix} \varphi(z) \\ z+1 \end{pmatrix}$$

$$t^{-1}st = \begin{pmatrix} \varphi(z) \\ \varphi(z+1) \end{pmatrix},$$

while

$$s^{\mu} = \begin{pmatrix} z \\ z+\mu \end{pmatrix}$$

can be written in the form

$$s^{\mu} = \begin{pmatrix} \varphi(z) \\ \varphi(z)+\mu \end{pmatrix}$$

because $\varphi(z)$, just as z, stands for the subscripts

$$1, 2, \ldots, p.$$

It follows that

$$\varphi(z+1) \equiv \varphi(z) + \mu (\text{mod } p).$$

Putting now

$$\varphi(0) \equiv \varphi(p) \equiv \nu,$$

we have

$$\varphi(1) \equiv \nu + \mu$$

$$\varphi(2) \equiv \varphi(1) + \mu \equiv \nu + 2\mu$$

$$\cdots\cdots$$

and in general:

$$\varphi(z) \equiv \nu + z\mu (\text{mod } p).$$

[1] Given in §1.

Hence it appears that

$$t = \begin{pmatrix} z \\ \mu z + \nu \end{pmatrix},$$

where

$$\mu = 1, 2, \ldots, p-1$$
$$\nu = 1, 2, \ldots, p-1, p.$$

Consequently there exist $p(p-1)$ distinct permutations t composing the group M, and we note:

(85) **The largest group of prime degree p containing the cyclic group of degree and order p as a normal subgroup is the metacyclic group of order $p(p-1)$ with all permutations of the form**

$$\boxed{t = \begin{pmatrix} z \\ \mu z + \nu \end{pmatrix}}.$$

The metacyclic group contains no circular permutations of order p other than those in C, for any permutation

$$t = \begin{pmatrix} z \\ \mu z + \nu \end{pmatrix}$$

with

$$\mu \neq 1$$

leaves unaltered that letter whose subscript is determined by the congruence[1]

$$\mu z + \nu \equiv z (\text{mod } p).$$

It will be shown in §81 that there exists a number g such that modulo p its powers

$$g, g^2, \ldots, g^{p-1}$$

are the numbers

$$1, 2, \ldots, p-1$$

in some order or other.

This makes it possible to represent the permutations of the metacyclic group in the form[2]

$$t = s^\lambda u^\mu,$$

[1] Cf. §80.

[2] $s^\lambda u^\mu = \begin{pmatrix} z \\ z+\lambda \end{pmatrix}\begin{pmatrix} z \\ g^\mu z \end{pmatrix} = \begin{pmatrix} z \\ g^\mu(z+\lambda) \end{pmatrix}$ stands for the same permutations as $t = \begin{pmatrix} z \\ \mu z + \nu \end{pmatrix}$.

where

$$s = \binom{z}{z+1}$$

$$u = \binom{z}{gz}.$$

A function ψ belonging to C is unaltered by s^λ while u converts

$$\psi \to \psi_u \to \psi_{u^2} \to \ . \ . \ . \ \to \psi_{u^{p-2}}.$$

All these functions are conjugate with ψ in M and belong to C which is normal.

Hence the group Γ of permutations between these functions, as caused by the permutations t of M, is cyclic, and so is the factor-group M/C.[1] This group thus has a normal subgroup of any index dividing $p - 1$, by proposition (75), consequently M has such a subgroup by proposition (68).

It follows that the metacyclic group is soluble; but we can prove even more:

(86) **The metacyclic group of prime degree is the largest transitive group of that degree which is soluble.**

For suppose that the metacyclic group M or any of its transitive subgroups is contained in a larger group G as a normal subgroup. If t is any permutation of G, it transforms the permutation s of M not into any permutation of M but a power of s, because the transform is by proposition (30) a circular permutation of order p, and there are no such permutations outside C.

Therefore

$$t^{-1}st = s^\mu,$$

and we are back to the relation which gave us the metacyclic group.

The metacyclic group of degree three is

$$\binom{z}{z} \quad \binom{z}{z+1} \quad \binom{z}{z+2} \quad \binom{z}{2z} \quad \binom{z}{2z+1} \quad \binom{z}{2z+2},$$

or respectively

$$1 \qquad (123) \qquad (132) \qquad (12) \qquad (13) \qquad (23),$$

which is the symmetric group as we should expect it for the degrees three and four.

[1] Cf. proposition (67).

The metacyclic group of degree five is:

$$\binom{z}{z} \quad \binom{z}{z+1} \binom{z}{z+2} \binom{z}{z+3} \binom{z}{z+4}$$

1, (12345), (13524), (14253), (15432)

$$\binom{z}{2z} \quad \binom{z}{2z+1}\binom{z}{2z+2}\binom{z}{2z+3}\binom{z}{2z+4}$$

(1243),(1325), (1452), (1534), (2354)

$$\binom{z}{3z} \quad \binom{z}{3z+1}\binom{z}{3z+2}\binom{z}{3z+3}\binom{z}{3z+4}$$

(1342),(1435), (1523), (2453), (1254)

$$\binom{z}{4z} \quad \binom{z}{4z+1}\binom{z}{4z+2}\binom{z}{4z+3}\binom{z}{4z+4}$$

(14)(23),(15)(24), (25)(34), (12)(35), (13)(45).

The permutations

$$t = s^{\lambda}v^{\xi},$$

where

$$s = \binom{z}{z+1}$$

$$v = \binom{z}{g^2z}$$

and

$$\xi = 1, 2, \ldots, \frac{p-1}{2},$$

compose a group which is called **half-metacyclic.**

The permutation

$$s = (12 \ldots p)$$

is even, the permutation

$$u = (1g \ldots g^{p-2})$$

is odd because it leaves x_p unaltered. The permutation v is even: it is composed of two equal cycles, since p $-$ 1 is divisible by 2 and g^{p-1} equals the subscript we started the cycle with.[1] Hence we infer for a prime degree that

(87) **the metacyclic group is not contained in the alternating group, but the half-metacyclic group is.**

No odd power of u can be an even permutation, because it has an odd number of odd cycles, u itself being an odd permutation.

[1] Cf. Fermat's Theorem in §81.

But an even power of u is a power of v; hence the half-metacyclic group contains all even permutations of the metacyclic group and

(88) **the half-metacyclic group is normal of index two in the metacyclic group**

by proposition (30).[1]

The half-metacyclic group of degree five is:

$$1, (12345), (13524), (14253), (15432),$$
$$(14)(23), (15)(24), (25)(34), (12)(35), (13)(45).$$

For this degree and

$$g = 2$$

we have

$$s = (12345), u = (1243), v = (14)(23).$$

A function unaltered by s and v is

$$\psi = x_1x_2 + x_2x_3 + x_3x_4 + x_4x_5 + x_5x_1.$$

It is converted by u into

$$\psi' = x_1x_3 + x_3x_5 + x_5x_2 + x_2x_4 + x_4x_1,$$

also unaltered by s and v. Both ψ and ψ' belong to the half-metacyclic group and are conjugate under the metacyclic group. The sum

$$\psi + \psi'$$

is symmetric, while the difference

$$\psi - \psi'$$

belongs to the half-metacyclic group and its square

$$\varphi = (\psi - \psi')^2$$

to the metacyclic group.

NOTE ON ABSTRACT GROUP

The concept of a group originated with the theory of permutations, but reaching out as it grew it finally became disassociated from any concrete operations whatsoever while applying to all of them. In its abstract form it embodies a certain law of combination exhibited on abstract operations or **elements,** and around its abstract form clusters the modern theory of **finite groups.**

Elements of a set, to qualify for the theory of groups, must

(1) possess a law of combination

(2) satisfy the associative law

[1] Because it is identical with its transforms.

(3) contain the **identical element**[1]

(4) contain the **inverse** of every element.

A set of elements forms a **group** if

(1) the product of any two elements is an element of the set

(2) the set contains the inverse of every element.

The number of elements in a group is its **order.**

Groups are **commutative** or **non-commutative** depending on whether their elements are commutative or non-commutative.

Groups can be **isomorphic,** but simply isomorphic groups are identical representing the same law of combination.

A group is defined by a set of **generating elements** satisfying independent and consistent relations. For instance, the generating elements a, b, c satisfying the relations

$$a^2 = 1, b^2 = 1, ab = ba = c$$

define a group of order four:

$$\Gamma = 1, a, b, c.$$

This we see from the multiplication table:

$$\begin{array}{c|ccc} 1 & a & b & c \\ \hline a & aa & ab & ac \\ b & ba & bb & bc \\ c & ca & cb & cc \end{array} = \begin{array}{c|ccc} 1 & a & b & c \\ \hline a & 1 & c & b \\ b & c & 1 & a \\ c & b & a & 1, \end{array}$$

for we obtain from the given independent relations the other ones:

$$ac = a^2b = b, \ cb = ab^2 = a$$
$$bc = b^2a = a, \ ca = ba^2 = b$$
$$cc = abba = 1.$$

The symmetry of the multiplication table around the main diagonal indicates that the group is commutative.

The theory of finite groups, developed from the theory of permutation-groups, remains essentially identical with it, for **Cayley's Theorem** states that

(89) **every group can be represented as permutation-group on its elements.**

Let the elements of a group Γ be

$$a, b, c, \ . \ . \ .$$

[1] Denoted by 1 if no confusion can arise; it is 0 in ordinary addition.

The combination of these elements with one of them, say a, obviously gives the same elements in the same or another order, so that

$$s_a = \begin{pmatrix} a & b & c & \cdots \\ aa & ba & ca & \cdots \end{pmatrix}$$

stands for a permutation. In the same sense we set

$$s_b = \begin{pmatrix} a & b & c & \cdots \\ ab & bb & cb & \cdots \end{pmatrix},$$

which can be written

$$s_b = \begin{pmatrix} aa & ba & ca & \cdots \\ aab & bab & cab & \cdots \end{pmatrix}$$

since the latter permutation, as the former, only replaces every element of Γ by the same element multiplied with b.

Combining s_a and s_b we obtain

$$s_a s_b = \begin{pmatrix} a & b & c & \cdots \\ aa & ba & ca & \cdots \end{pmatrix}\begin{pmatrix} aa & ba & ca & \cdots \\ aab & bab & cab & \cdots \end{pmatrix} = \begin{pmatrix} a & b & c & \cdots \\ aab & bab & cab & \cdots \end{pmatrix} = s_{ab}$$

and see that the permutations s_a and s_b combine as the elements a and b do. Hence it follows that the groups

$$\begin{aligned} \Gamma &= a, \quad b, \quad c, \quad \ldots \\ G &= s_a, \quad s_b, \quad s_c, \quad \ldots \end{aligned}$$

are simply isomorphic and abstractly speaking identical, which proves the proposition.

For our example of an abstract group we find

$$\begin{aligned} \Gamma &= 1, \quad a, \quad b, \quad c \\ G &= 1, (1a)(bc), (1b)(ac), (1c)(ab) = V. \end{aligned}$$

CHAPTER X

DOMAIN

§59. ALGEBRAIC DOMAIN

Numbers compose a **domain or field** if they combine into such numbers only as are among them when acted upon by the rational operations of arithmetic (except division by zero).

Acting upon numbers 1, the rational operations produce the domain formed by the rational numbers of arithmetic. It is commonly called the **domain of rational numbers** and denoted by (1).

Every domain[1] contains the number 1 as quotient of any number other than zero in the domain by itself, and contains therefore the domain (1) of rational numbers.

If we **adjoin** to the domain (1) a number α not contained in it, we obtain a larger domain denoted by (α); it includes (1) and α and everything that the rational operations can give acting upon α and the numbers of (1). If we adjoin to the domain (α) a number β not contained in it, we obtain a still larger domain (α, β).

Whenever an implicit notation is expedient, we denote a domain by Ω, a symbol which qualifies for this use by its very form, it would seem. A number in Ω we can denote by ω. If we set

$$\Omega = (\alpha),$$

then

$$\Omega(\beta) = (\alpha, \beta).$$

Every number belongs to some domain. An integral function

$$f(x) = a_0x^n + a_1x^{n-1} + \dots + a_n$$

is considered as belonging to such a domain

$$(a_i)_1^n = (a_1, \dots, a_n)$$

as contains its coefficients and hence symmetric functions of its roots. Number and function are said to be **rational in a domain**

[1] Other than (0), which we do not regard as forming a domain properly speaking.

to which they belong, and a number also to be **rationally known** there.[1] It may be well to note that we mean by a domain to which a number or function belongs the smallest such domain, unless specified otherwise.

When we say in elementary algebra that a function $x^2 - 2$ is irreducible, we do not quite mean what we say: we only mean to say that it is irreducible in the domain (1) of rational numbers, for evidently the function

$$x^2 - 2 = (x + \sqrt{2})(x - \sqrt{2})$$

is reducible in the domain $(\sqrt{2})$. And although the function

$$x^2 + 2 = (x + \sqrt{2}\, i)(x - \sqrt{2}\, i)$$

is irreducible in the domains (1) and $(\sqrt{2})$, it is reducible in the domain $(\sqrt{2}, i)$.

Thus, a function is reducible or irreducible only with reference to a domain. Any function becomes reducible if we adjoin a root of the function to its domain, and the linear function alone is not reducible any further.

We shall denote functions reducible in a domain preferably by large letters and functions irreducible so preferably by small letters.

An equation

$$F(x) = 0$$

is considered rational and reducible in the domain in which the function $F(x)$ is so.

If an equation

$$f(x) = a_0x^n + a_1x^{n-1} + \ldots + a_n = 0$$

is rational but irreducible in the domain

$$\Omega = (a_i)_1^n,$$

it becomes reducible in the domain

$$\Omega(x_1) = (a_i, x_1)_1^n$$

which we obtain adjoining to $(a_i)_1^n$ the root x_1 of the equation; for there we can separate from $f(x)$ the factor $x - x_1$. Hence the equation can be solved in the domain

$$\Omega(x_i)_1^n = (a_i, x_i)_1^n,$$

where it is reducible to linear factors.

[1] In short, we shall say that a number or function is in Ω, instead of rational in Ω.

Domains that we obtain by adjoining to the domain Ω of an equation its algebraic roots[1]—which alone concern us here—are called **algebraic domains** on Ω, and Ω is said to be a **subdomain** of those. Not every equation is reducible in an algebraic domain since not every equation has algebraic roots.[2]

For integral functions we note:

(90) **A function $f(x)$ which is rational but irreducible in the domain Ω has no common factor with another function $F(x)$ which is rational in Ω unless the entire function $f(x)$ is a factor of the function $F(x)$.**

This is obvious since the highest common factor of the functions $F(x)$ and $f(x)$ is computable from them by rational operations[3] and therefore belongs to their domain Ω, while we assume that the function $f(x)$ has no factor rational in Ω other than itself.

It follows that the function $f(x)$ has none of its roots or all of them in common with the function $F(x)$, and we note in particular that it has all of them in common if it has one.

It also follows that an irreducible function never has a double root, since it cannot have a root in common with its derivative.

§60. ALGEBRAIC DOMAIN, *Continued*

A root x_1 of an equation

$$F(x) = 0$$

rational[4] in the domain Ω is also a root of an irreducible equation

$$f(x) = 0$$

rational in Ω. For $F(x)$, if not itself irreducible in Ω, must have such a factor $f(x)$ containing the root x_1. If we set

$$f(x) = x^n + a_1x^{n-1} + \cdots + a_n$$

with

$$a_0 = 1,$$

then x_1 is a root of a very definite equation rational and irreducible in Ω.

It is the degree of this equation which determines the **degree** of the algebraic domain $\Omega(x_1)$.

[1] We mean algebraically computable roots. Cf. §1.

[2] Cf. §82, end.

[3] By proposition (2).

[4] This refers to the equation.

We now prove for any root x_1 that

(91) **the algebraic domain $\Omega(x_1)$ consists of all integral functions of x_1 which are rational in Ω.**

Since any number ω_1 in $\Omega(x_1)$ can be obtained from x_1 and the numbers in Ω by rational operations, it is expressible as function of x_1 rational in Ω:

$$\omega_1 = \frac{\varphi(x_1)}{\psi(x_1)},$$

where φ and ψ are integral functions. Since $\psi(x_1)$ cannot be zero, $\psi(x)$ has no factor in common with $f(x)$ which is irreducible,[1] and by proposition (4) we may set

$$Rf + r\psi = \varphi,$$

where R and r are integral functions rational in Ω and r is of degree less than n.

If now

$$x = x_1,$$

then

$$f = 0$$

and

$$\frac{\varphi(x_1)}{\psi(x_1)} = r(x_1).$$

Hence any number in $\Omega(x_1)$ is expressible as integral rational function of x_1:

$$\omega_1 = r(x_1).$$

This function is unique as from another relation

$$\omega_1 = \rho(x_1)$$

would follow that the equation

$$r(x) - \rho(x) = 0$$

of degree less than n and rational in Ω is satisfied by

$$x = x_1.$$

But then all other roots of $f(x)$ would have to satisfy this equation by proposition (90), for $f(x)$ is rational and irreducible in Ω. This is impossible since $f(x)$ is of degree n, which completes our proof.

[1] By proposition (90).

§61. CONJUGATE DOMAINS

No root x_1 of the irreducible function $f(x)$ of degree n in the domain Ω is rational in Ω, since the function $f(x)$ has no factor $x - x_1$ rational in Ω. Adjoining the different roots x_i of $f(x)$ to Ω, we obtain the **conjugate domains** on Ω:

$$\Omega(x_1), \Omega(x_2), \ldots, \Omega(x_n),$$

and we note that they all are of degree n. The number of conjugate domains thus equals the degree of any one.

Such numbers in the conjugate domains as we obtain by giving to the variable x in the function $r(x)$ the values x_i:

$$\omega_1 = r(x_1), \omega_2 = r(x_2), \ldots, \omega_n = r(x_n),$$

are called **conjugate numbers;** there is one in each conjugate domain.

It is clear that

(92) **every number ω_1 in an algebraic domain of degree n on Ω is a root of an equation**

$$\boxed{\Phi(\omega) = (\omega - \omega_1)(\omega - \omega_2) \ldots (\omega - \omega_n) = 0}$$

of degree n whose other roots are the numbers conjugate with ω_1 and which is rational in Ω.

For the coefficients of $\Phi(\omega)$ are symmetric in the x_i since any permutation on the x_i only interchanges the ω_i.

If $\Phi(\omega)$ is reducible in Ω, as it may be, it must have there irreducible factors since the ω_i are not in Ω. If $\varphi(\omega)$ is one such factor, the equation

$$\varphi(\omega) = 0$$

in Ω must be satisfied by some number ω_i, say ω_1. From

$$\varphi(\omega_1) = 0,$$

which is the same as

$$\varphi[r(x_1)] = 0,$$

it then appears that x_1 is a root of the equation

$$\varphi[r(x)] = 0$$

in Ω. Since a root of the irreducible equation

$$f(x) = 0$$

in Ω cannot satisfy another equation of the same domain unless all its roots do so,[1] we then have

$$\varphi[r(x_i)] = 0, \qquad [i = 1, \ldots, n$$

which is the same as

$$\varphi(\omega_i) = 0.$$

Hence $\varphi(\omega)$ has all the roots of $\Phi(\omega)$ and consequently

$$\Phi(\omega) = \varphi(\omega).$$

But this is completely true only if all numbers ω_i are distinct. If among them the numbers

$$\omega_1, \ldots, \omega_m$$

alone are distinct while any other number ω_i equals one of these, we still assume that the function

$$\varphi(\omega) = (\omega - \omega_1) \ldots (\omega - \omega_m)$$

is rational and irreducible in Ω; but the function $\Phi(\omega)$ then is reducible in Ω since it has the factor $\varphi(\omega)$. Any other irreducible factor of $\Phi(\omega)$ must be equal to $\varphi(\omega)$,[1] having some root ω_i in common with $\varphi(\omega)$, whence

$$\Phi(\omega) = \varphi^k(\omega),$$

where

$$km = n.$$

Thus,

(93) **the function $\Phi(\omega)$ is either irreducible in Ω or else some power of an irreducible function, while conjugate numbers ω_i are either all unlike or else divide into equal sets of unlike numbers.**

A number of an algebraic domain $\Omega(x_1)$ different from its conjugates is called a **primitive number** of that domain and may be denoted by θ_1. We observe that it is a root of an irreducible equation in Ω which has the same degree as the domain $\Omega(x_1)$ has on Ω.

If the numbers ω_i are **imprimitive**, dividing into equal sets of m unlike numbers, a number ω_1 evidently is primitive in an algebraic domain of degree m on Ω; whence it is rational in Ω if the ω_i are all alike.

[1] By proposition (90).

§62. CONJUGATE DOMAINS, *Continued*

Let ω_1 be any number in the algebraic domain $\Omega(x_1)$ of which θ_1 is a primitive number, and let their conjugate values be

$$\omega_1 = r(x_1),\ \omega_2 = r(x_2),\ .\ .\ .\ ,\ \omega_n = r(x_n)$$
$$\theta_1 = \rho(x_1),\ \theta_2 = \rho(x_2),\ .\ .\ .\ ,\ \theta_n = \rho(x_n).$$

The function

$$\frac{\omega_1}{\theta - \theta_1} + \frac{\omega_2}{\theta - \theta_2} + \ .\ .\ .\ + \frac{\omega_n}{\theta - \theta_n}$$

of θ is rational in Ω since it is symmetric in the x_i, any permutation on the x_i only interchanging its terms. Multiplying it by the function

$$\varphi(\theta) = (\theta - \theta_1)(\theta - \theta_2)\ .\ .\ .\ (\theta - \theta_n)$$

which also is rational in Ω, we obtain

$$\varphi(\theta)\left(\frac{\omega_1}{\theta - \theta_1} + \frac{\omega_2}{\theta - \theta_2} + \ .\ .\ .\ + \frac{\omega_n}{\theta - \theta_n}\right) = \chi(\theta)$$

which is an integral function rational in Ω and of degree $n - 1$ in θ if we expand it.

Setting in this function

$$\theta = \theta_1,$$

we find that

$$\omega_1 = \frac{\chi(\theta_1)}{(\theta_1 - \theta_2)\ .\ .\ .\ (\theta_1 - \theta_n)} = \frac{\chi(\theta_1)}{\varphi'(\theta_1)},$$

where $\varphi'(\theta)$ is the derivative of $\varphi(\theta)$ and as such an integral function rational in Ω; $\varphi'(\theta_1)$ cannot be zero since the θ_i are all unlike.

Thus we proved that

(94) **every number ω_1 of an algebraic domain $\Omega(x_1)$ is expressible as a function rational in Ω of any primitive number θ_1 in $\Omega(x_1)$:**

$$\boxed{\omega_1 = R(\theta_1)}.$$

This obviously means that the domains $\Omega(\theta_1)$ and $\Omega(x_1)$ are identical:

$$\Omega(\theta_1) = \Omega(x_1).$$

The primitive number θ_1 is a root of an irreducible equation in Ω of degree n while the imprimitive number ω_1 is a root of an irreducible equation in Ω of degree less than n.[1] Hence θ_1

[1] By proposition (93); cf. §61, end.

cannot be rational in $\Omega(\omega_1)$ and the latter is a subdomain of $\Omega(\theta_1)$ or $\Omega(x_1)$:

$$\Omega(\omega_1) < \Omega(x_1).$$

An algebraic domain on Ω containing no imprimitive numbers[1] is called a **primitive domain**; it certainly is primitive if its degree is prime, since its numbers then cannot divide into sets. When not primitive, an algebraic domain is called **imprimitive.** A primitive domain on Ω contains no other subdomain than Ω.

Proposition (94) seems to recall the Theorem of Lagrange: there, functions of one group rationally expressible in terms of each other; here, functions of one domain expressible so. We feel that the circle of our investigation is going to close.[2]

§63. NORMAL DOMAIN

If an algebraic domain of an irreducible equation is identical with its conjugate domains, it is called a **normal or invariant domain,** and the equation is called a **normal or invariant equation.** It appears that a normal equation is irreducible and has roots which are rationally expressible in terms of each other.

Suppose that the equation

$$f(x) = 0$$

of degree n, rational and irreducible in Ω, is normal; then

$$\Omega(x_1) = \Omega(x_2) = \ldots = \Omega(x_n).$$

Let θ_1 be a primitive number in $\Omega(x_1)$ and have the conjugate values

$$\theta_1, \theta_2, \ldots, \theta_n.$$

Since

$$\Omega(\theta_i) = \Omega(x_i)$$

by proposition (94), we have

$$\Omega(\theta_1) = \Omega(\theta_2) = \ldots = \Omega(\theta_n),$$

and consequently the θ_i are roots of the equation

$$g(\theta) = 0$$

also normal of degree n in Ω.

[1] Outside those in Ω, of course.

[2] The proof of proposition (94) can be applied to Lagrange's Theorem, and the proof given there can be applied here; but the present proof adds an elegant tool to the mathematical outfit of the student. Compare §69.

Any one root of a normal equation in Ω, and of such an equation alone, is expressible as a function rational in Ω of any other root and we may set

$$\theta_1 = R_1(\theta_1),\ \theta_2 = R_2(\theta_1),\ \ldots,\ \theta_n = R_n(\theta_1).$$

If on these functions we perform a **substitution**[1] replacing θ_1 by θ_k, we obtain as result the same functions in another order because from

$$g[R_i(\theta_1)] = 0$$

follows[2] that every root of the irreducible equation

$$g(\theta) = 0$$

in Ω satisfies the equation

$$g[R_i(\theta)] = 0$$

in Ω, whence with

$$R_i(\theta_1)$$

also

$$R_i(\theta_k)$$

is a root of

$$g(\theta) = 0.$$

And no two roots $R_i(\theta_k)$ and $R_j(\theta_k)$ are alike since from

$$R_i(\theta_k) - R_j(\theta_k) = 0$$

would follow as before that also

$$R_i(\theta_1) - R_j(\theta_1) = 0,$$

which is untrue.

The substitution that we performed on the θ_i we denote by

$$\sigma_k = (\theta_1\theta_k).$$

It converts any root

$$\theta_i = R_i(\theta_1)$$

into a root θ_j defined by the relation

$$\theta_j = R_i(\theta_k) = R_i[R_k(\theta_1)] = R_j(\theta_1),$$

so that of two roots like θ_i and θ_j we may choose one at pleasure while the other is fixed by the substitution.

Any number

$$\omega_1 = \varphi(\theta_1)$$

rational in $\Omega(\theta_1)$ is converted by the substitution σ_k into

$$\omega_k = \varphi(\theta_k).$$

[1] Compare §9.

[2] By proposition (90).

As ω_1 is rational also in $\Omega(\theta_i)$, we may set

$$\omega_1 = \psi(\theta_i),$$

and this is converted by σ_k into

$$\omega_k = \psi(\theta_j),$$

whence

$$\sigma_k = (\theta_1\theta_k) = (\theta_i\theta_j).$$

The substitutions $(\theta_i\theta_j)$ are called **substitutions of the normal domain** $\Omega(\theta_1)$, and since they can be expressed in the form $(\theta_1\theta_k)$, the normal domain has as many different substitutions as its degree indicates, which is n.

Every one of these substitutions converts a number ω_1 of the normal domain $\Omega(\theta_1)$ into a number of the same domain, and no two numbers into the same number. If a number ω_1 remains unaltered by every substitution of the normal domain, it is identical with its conjugates and therefore rational in Ω.

The substitutions of the normal domain $\Omega(\theta_1)$ are qualified elements composing a group in the sense of the note to chapter nine. For

$$\sigma_i\sigma_k = \sigma_j$$

since

$$(\theta_1\theta_i)(\theta_i\theta_j) = (\theta_1\theta_j),$$

if we arrange it so that the first letter in the second substitution is identical with the second letter in the first substitution, and the other qualifications are readily verified.

It appears that

(95) **the substitutions of a normal domain compose a group**

$$\boxed{\langle\Gamma\rangle = (\theta_1\theta_i)_1^n},$$

and this leads us back to the theory of equations.

CHAPTER XI

THEORY OF GALOIS

§64. SPECIAL EQUATION

Disappointed in our effort to solve the general equation, we now turn to special equations, that is to say equations with numerical coefficients. Therefore it has to be kept in mind firmly that in all what follows we are dealing with numerical values, unless specified otherwise.

Lagrange's plan of solving the general equation falls short if applied to numerical equations whenever we strike conjugate functions that are equal. Thus, for the equation

$$x^3 - 2 = 0$$

with the roots

$$x_1 = 2^{1/3},\ x_2 = \omega 2^{1/3},\ x_3 = \omega^2 2^{1/3}$$

the conjugate functions

$$\begin{aligned} \psi_1 &= x_2 x_3 - x_1^2 = 0 \\ \psi_2 &= x_1 x_3 - x_2^2 = 0 \\ \psi_3 &= x_1 x_2 - x_3^2 = 0 \end{aligned}$$

would not permit a computation by the formula

$$\varphi_1 = \frac{R(\psi_1)}{\Delta_\psi}$$

of Lagrange's Theorem, since the denominator becomes zero. Likewise, for the equation

$$x^4 + 1 = 0$$

with the roots

$$x_1 = i^{1/2},\ x_2 = i^{3/2},\ x_3 = i^{5/2}\ x_4 = i^{7/2}$$

the conjugate functions

$$\begin{aligned} \chi_1 = x_1^2 \quad & \chi_2 = x_2^2 \quad & \chi_3 = x^{\ 2} \quad & \chi_4 = \chi_4^2 \\ = i, \quad & \quad = -i, \quad & = i, \quad & \quad = -i. \end{aligned}$$

would not permit such a computation.

Furthermore, the permutations that leave ψ_1 unaltered in value, either because they leave it unaltered in formal value:

$$1, (23),$$

or because they convert it into ψ_2 and ψ_3:

$$(12), (13), (123), (132),$$

they do form a group. But those that leave χ_1 unaltered in value, either because they leave it unaltered in formal value:

$$1, (23), (24), (34)\ (234), (243),$$

or because they convert it into χ_3:

$$(13), (13)(24), (132), (134), (1324), (1342),$$

they do not form a group. Hence, passing to numerical values, we cannot say any longer that permutations which leave the value of a function unaltered compose a group.

To become applicable to special equations, Lagrange's plan had to be modified, and this work was done by Galois.[1] Lagrange's special plan of solving general equations young Galois replaced by his general plan of solving special equations, and the light touch of genius marks his achievement: it is an epic of human glory

"not ill-befitting men that strove with gods."

§65. GALOISIAN FUNCTION

Permutations that leave a function numerically unaltered do not in general form a group; but when they concur with those that leave the function also formally unaltered, they evidently do so. Discarding functions that remain numerically unaltered by permutations which do not leave them formally unaltered, Galois taught us to construct functions that belong to a group both formally and numerically; and such functions we call **Galoisian functions.**

Let

$$F(x) = a_0x^n + a_1x^{n-1} + \ . \ . \ . + a_n = 0$$

be a special equation rational in the domain

$$\Omega = (a_i)_1^n$$

[1] Galois lived 1811–1832. He was seventeen when he obtained his first beautiful results in mathematics, at nineteen he had divined the nature of equations, and he was twenty when a duel ended his stormy life.

of its coefficients, or any wider domain Ω, and let us impose on it the condition that it has no double roots. Evidently this condition does not limit the range of the theory which is to be built up, for double roots of a function are intimated by its vanishing discriminant and readily found and eliminated as roots of the common factor which the function has with its derivative.

Then,

(96) **it is always possible to construct in the domain Ω of an equation a Galoisian function of its roots x_i which belongs to the group identity on the x_i.**

Altered by every permutation on the x_i both formally and numerically, such a function v_1 takes under the symmetric group

$$S = 1, s_2, \ldots, s_{n!}$$

on the x_i the conjugate values

$$v_1, v_2, \ldots, v_{n!}$$

and may be called an **elementary Galoisian function** for the given equation.

Galois puts

$$v_1 = u_1x_1 + u_2x_2 + \ldots + u_nx_n$$

with indeterminate u_i; for the integral rational function[1]

$$\prod_{j \lessgtr k}(v_j - v_k) = J(a_i, u_i)_1^n \quad [j, k = 1, \ldots, n!$$

does not vanish identically,[2] so that we may assign to the u_i integral numerical values that do not make $\prod$ vanish, that is to say do not make two values v_j equal.

It follows that

(97) **it is always possible to construct in the domain Ω of an equation a Galoisian function of its roots x_i which belongs to any group on the x_i,**

both formally and numerically.

Suppose such a group is

$$G = 1, s_2, \ldots, s_r$$

converting the elementary Galoisian function v_1 into

$$v_1, v_2, \ldots, v_r,$$

[1] Integral and rational in both the a_i and the u_i, for symmetric in the x_i.

[2] Because the v_j are distinct by §16, the u_i being indeterminate quantities.

so that like subscripts identify the conjugate values into which the permutations of G convert v_1. Galois obtains a function which satisfies the proposition by putting[1]

$$g(v) = (v - v_1)(v - v_2) \ldots (v - v_r),$$

where v is a rational number properly to be chosen but for the time indeterminate.

For a permutation s_k of G only interchanges the v_j, since applied to v_j it gives the same result as $s_j s_k$ applied to v_1 and hence a conjugate value which is in the set as $s_j s_k$ is in G. While a permutation of G thus leaves $g(v)$ unaltered, a permutation not in G converts the v_j of the set into such as are outside and alters $g(v)$.

It will be observed that functions as Galois taught us to construct are integral functions.

Applying Lagrange's Theorem to Galoisian functions, we evidently need not fear trouble:

(98) **If a rational function φ in the roots x_i of a special equation remains formally unaltered by all those permutations on the x_i that leave a Galoisian function g of the x_i unaltered, then the function φ is rationally expressible in terms of the function g;**

here as rational is considered the domain Ω of the equation.

For the proposition is true by virtue of Lagrange's Theorem if we take the x_i to be indeterminate quantities:

$$\varphi = \frac{R(g)}{\Delta_g}.$$

On replacing the indeterminate x_i by numerical values we meet no difficulty, since the discriminant in the denominator cannot be zero.

In particular, any function φ is rationally expressible in terms of the elementary Galoisian function v_1:

$$\varphi(x_i) = R(v_1).$$

As this relation is an identity in the x_i, both members remain identical when we operate a permutation s on the x_i. This gives

$$\varphi_s(x_i) = R(v_s),$$

[1] To express that $g(v)$ is a function also of the x_i we denote it by $g(v|x_i)$.

the subscript s identifying the permutation that converts φ_1 into φ_s and v_1 into v_s. Hence we note:

(99) **If $\varphi(x_i)$ is a rational function of v_1, then $\varphi_s(x_i)$ is the same function of v_s, the v_i being elementary Galoisian functions in the x_i and s a permutation on the x_i.**

The proposition holds true for the roots x_i themselves:

$$x_1 = R_1(v_1),\ x_2 = R_2\ (v_1),\ .\ .\ .\ ,\ x_n = R_n(v_1),$$

or

$$x_1 = R_1(v_1),\ x_2 = R_1(v_2),\ .\ .\ .\ ,\ x_n = R_1(v_n)$$

when for instance

$$s_2 = (12),\ .\ .\ .\ ,\ s_n = (1n).$$

This recalls to our mind that a special equation is solved when we know a Galoisian function v_1 of its roots belonging to identity.

§66. GALOISIAN RESOLVENT

The function

$$G(v) = (v - v_1)(v - v_2)\ .\ .\ .\ (v - v_{n!})$$

which belongs to the symmetric group on the x_i we call a **complete Galoisian function** for the special equation

$$F(x) = 0.$$

It may happen that such a complete Galoisian function $G(v)$ is reducible in the domain Ω of the equation—which evidently implies that there exist functions[1] of the x_i not symmetric and yet rational in Ω. If $G(v)$ is reducible, it must have an irreducible factor $g(v)$ since we assume that the equation cannot be solved in Ω; for when it can, there is no problem left to be considered.

The function $g(v)$, still rational in Ω but not further reducible there, must vanish for at least one value v_i, say v_1. If it has more roots, let them be

$$v_1,\ v_s,\ .\ .\ .\ ,\ v_t,$$

subscripts again identifying the permutations on the x_i which applied to v_1 give its conjugate values. The function

$$g(v) = (v - v_1)(v - v_s)\ .\ .\ .\ (v - v_t)$$

is called a **primary Galoisian function** and the equation

$$g(v) = 0$$

a **Galoisian resolvent** for the given equation.

[1] But they can exist even though $G(v)$ be irreducible, as will appear in §70.

If the complete Galoisian function $G(v)$ is irreducible in Ω, it is itself a primary Galoisian function and

$$G(v) = 0$$

a Galoisian resolvent. Furthermore it is to be noted that also an equation

$$g(\psi) = 0$$

often is called a Galoisian resolvent, when ψ is any function numerically belonging to identity, not necessarily a Galoisian function belonging to identity both formally and numerically.

For a normal equation we may set

$$v_i = x_i,$$

because a normal equation is irreducible and its roots x_i belong to identity being functions rational in Ω of each other;[1] identity is the only permutation which the groups of the x_i can have in common.

Again, since all roots v_i of a Galoisian resolvent belong to identity, they are expressible by proposition (98) as functions rational in Ω of v_1. This permits to conclude that

(100) **every Galoisian resolvent is a normal equation, and every normal equation is its own Galoisian resolvent;**

further that

(101) **the domain $\Omega(v_1)$ is a normal domain if v_1 is an elementary Galoisian function for a special equation in Ω.**

Also the roots x_i are expressible as functions rational in Ω of v_1, while we have

$$v_1 = u_1x_1 + u_2x_2 + \dots + u_nx_n.$$

Hence it appears that

$$\Omega(v_1) = \Omega(x_i)_1^n,$$

which is to say that

(102) **the adjunction of an elementary Galoisian function for a special equation is equivalent to the adjunction of all its roots.**

Thus Galois has taught us to construct a normal equation for any given equation and for its domain an algebraic domain which is normal.

[1] Being rational functions of each other, they must belong to the same group, by proposition (98) which presents a necessary and sufficient condition; compare §36.

§67. GALOISIAN GROUP

The permutations on the x_i converting a root v_1 of

$$G(v) = (v - v_1)(v - v_2) \ldots (v - v_{n!}) = 0$$

into all its roots evidently compose a group. Galois discovered this always to be true for his resolvent:

(103) **The permutations on the x_i converting a root v_1 of the Galoisian resolvent**

$$\boxed{g(v) = (v - v_1)(v - v_s) \ldots (v - v_t) = 0}$$

into all its roots compose a group.

We call it the **Galoisian group** or simply **group of the equation**

$$F(x) = 0$$

in Ω and denote it by G or $\langle G \rangle$.

Consider the root v_s of the Galoisian resolvent. It is expressible as function rational in Ω of the root v_1:

$$v_s = R_s(v_1),$$

so that

$$g(v_s) = g[R_s(v_1)] = 0.$$

But also

$$g(v_1) = 0,$$

whence v_1 is a root of two equations in Ω:

$$g(v) = 0$$

and

$$g[R_s(v)] = 0.$$

Since the first equation is irreducible in Ω, all its roots must satisfy the second equation as one does.[1] In particular, the root v_t must do so; this gives

$$g[R_s(v_t)] = 0,$$

whence $R_s(v_t)$ is a root of

$$g(v) = 0.$$

To find the significance of this root, we apply to the identity

$$v_s = R_s(v_1)$$

in the x_i the permutation t and obtain[2]

$$v_{st} = R_s(v_t).$$

[1] By proposition (90).
[2] By proposition (99).

Hence it appears that v_{st} is among the roots of the Galoisian resolvent and therefore st among the permutations converting v_1 into those roots.

This proves that the permutations on the x_i converting v_1 into

$$v_1, \quad v_s, \quad . \; . \; . \; , v_t$$

compose a group, the Galoisian group

$$G = 1, \quad s, \quad . \; . \; . \; , t$$

of the given equation. Its order is equal to the degree of the Galoisian resolvent for that equation.

It is clear that the primary Galoisian function $g(v)$ belongs to the group G of the equation; for the permutations of G do nothing more than interchange its roots, and they are the only permutations which do so.

Since any two roots v_i of the Galoisian resolvent are interchanged by just one permutation of the Galoisian group[1] and just one substitution of the normal domain, we obtain a beautiful result:

(104) **The Galoisian group of a special equation and the substitution group of its normal domain are simply isomorphic and abstractly identical:**

$$\boxed{\langle G \rangle = \langle \Gamma \rangle}.$$

This links up the theory of groups with the theory of domains.

§68. PROPERTIES OF GALOISIAN GROUP

The Galoisian group of the special equation

$$F(x) = 0$$

rational in the domain of its coefficients or any wider domain has two fundamental properties which we formulate referring rationality to the domain Ω of the equation.

(105) **Property (1): Every rational function of the x_i which has a rational value remains numerically unaltered under the Galoisian group.**

For let such a function be

$$\varphi_1(x_i) = \omega,$$

[1] Only the permutation s converts v_1 into v_s, for instance.

where ω is a number rational in Ω. Expressing the function rationally in terms of

$$v_1 = u_1x_1 + u_2x_2 + \ldots + u_nx_n,$$

which we can do by proposition (98), we set

$$\varphi_1 = R(v_1).$$

Now we apply to this identity in the x_i the permutations of the Galoisian group and obtain[1]

$$\varphi_s = R(v_s)$$

$$\cdots\cdots\cdots$$

$$\varphi_t = R(v_t).$$

Since the rational equation

$$R(v) = \omega$$

is satisfied by the root v_1 of the rational and irreducible resolvent

$$g(v) = 0,$$

it is satisfied by every root of it. Therefore we have

$$R(v_1) = R(v_s) = \ldots = R(v_t) = \omega$$

and consequently

$$\varphi_1 = \varphi_s = \ldots = \varphi_t = \omega,$$

which proves the proposition.

If a permutation s' leaves every rational function of the x_i which has a rational value numerically unaltered, then with

$$g(v_1) = 0$$

which we know to be true also

$$g(v_{s'}) = 0.$$

Hence $v_{s'}$ is a root of the Galoisian resolvent and consequently s' a permutation of the Galoisian group. This means that

(106) **the Galoisian group is the largest group under which every rational function of the x_i which has a rational value remains numerically unaltered.**

Since a rational equation

$$\varphi(x_i) = \omega$$

can be written in the form

$$\chi(x_i) = \varphi(x_i) - \omega = 0,$$

[1] By proposition (99).

so that the function $\chi(x_i)$ has a value rational in any domain, it follows from property (1) that

(107) **every rational equation in the x_i remains true when operated on by the permutations of the Galoisian group;**

or permits those permutations, as we say.

The other property is

(108) **Property (2): Every rational function of the x_i which remains numerically unaltered under the Galoisian group has a rational value.**

Assuming that

$$\varphi_1 = \varphi_s = \ldots = \varphi_t,$$

we can set

$$\varphi_1 = \frac{1}{r}[R(v_1) + R(v_s) + \ldots + R(v_t)],$$

if r denotes the order of the Galoisian group. Since the coefficients of the Galoisian resolvent have rational values and every symmetric function of its roots is rationally expressible in terms of its coefficients, the symmetric function

$$R(v_1) + R(v_s) + \ldots + R(v_t)$$

of its roots has a rational value, which proves the proposition.

If some group

$$G' = 1, s', \ldots, t'$$

possesses property (2), the coefficients of the function

$$g'(v) = (v - v_1)(v - v_{s'}) \ldots (v - v_{t'})$$

have rational values by assumption, for they remain numerically unaltered under G'. But being rational and having a root v_1 in common with the rational and irreducible function $g(v)$, the function $g'(v)$ must contain all its roots and consequently G' all the permutations of the Galoisian group. This means that

(109) **the Galoisian group is the smallest group such that every rational function of the x_i which remains under it numerically unaltered has a rational value.**

Since the Galoisian group is the largest group possessing property (1) and the smallest group possessing property (2), it follows that

(110) **the Galoisian group of an equation is uniquely defined by its fundamental properties.**

Now it is clear that the group of an equation does not depend on our choice of v_1 among the roots of the complete Galoisian function: any irreducible factor of

$$G(v) = g(v) \cdot g'(v) \ldots$$

may serve as primary Galoisian function and must yield the same group. The group of the equation does not even depend on the construction of v_1 which is possible in any number of ways.

§69. PLAN OF GALOIS

The coefficients of a Galoisian resolvent for a special equation are rationally known while any one root solves the equation. Therefore the plan for solving a special equation depends on the composition-series not of the symmetric but of the Galoisian group for the equation, as the coefficients of the resolvent belong to that group and every root belongs to identity.

The Galoisian group takes in the theory of special equations the same place which the symmetric group has in the theory of general equations, and the permutations of the symmetric group that are outside the Galoisian group are completely ignored in the theory of special equations.

In this theory we can make use of any rational functions remaining numerically unaltered by the permutations of a group, they do not have to be Galoisian; but these establish the possibility of constructing such functions and link up the plan of Galois with the plan of Lagrange.

Let G be the Galoisian group of the special equation

$$F(x) = 0$$

rational in Ω. In the theory of special equations a function of the roots x_i is said to belong to a subgroup H of G when it remains numerically unaltered by all those, and only those, permutations of G which are in H; so that also in the theory of special equations

(111) **to every group on the x_i belongs a rational function of the x_i.**

Conversely, the permutations that leave a rational function unaltered compose a group, so that also in the theory of special equations

(112) **every rational function of the x_i belongs to a group on the x_i.**

To prove this, let s and s' be two permutations of H leaving a rational function ψ_1 of the x_i unaltered, whence

$$\psi_1 = \psi_s = \psi_{s'}.$$

Then with

$$\psi_{s'} = \psi_1$$

also

$$\psi_{s's} = \psi_s,$$

by proposition (107). But from

$$\psi_{s's} = \psi_s = \psi_1$$

follows that the permutation $s's$ leaves the function ψ_1 unaltered; consequently it is contained in the same group H with the permutations s and s'.

Also in the theory of special equations the function ψ_1 takes under the partitions

$$H, Ht, \ldots$$

of G the conjugate values

$$\psi_1, \psi_t, \ldots$$

For with

$$\psi_s = \psi_1$$

also

$$\psi_{st} = \psi_t,$$

by proposition (107). Hence all propositions in the theory of general equations which follow from this are readily verified for special equations. In particular:

(113) **If a group has a subgroup of index j, a rational function belonging to the subgroup takes j conjugate values under the group which belong to conjugate subgroups.**

Also in the theory of special equations

(114) **the conjugate values of a rational function under a group to whose subgroup the function belongs are roots of a resolvent equation whose degree equals the index of the subgroup in the group and whose coefficients are not altered by the group.**

If the resolvent is

$$r(\psi) = (\psi - \psi_1) \ldots (\psi - \psi_t) = 0,$$

a permutation t' of the group only interchanges its roots since any permutation tt' is contained in a partition of the group.

Conjugate values ψ_i of a function are roots of a resolvent equation in Lagrange's sense of the term, an ordinary resolvent so to say, well to be distinguished from a Galoisian resolvent.

Any rational function of the x_i which belongs to the group G is a number in Ω by property (2) of G. It follows for the domain Ω that

(115) **the conjugate values which a rational function takes under the group of a special equation are roots of a rational and irreducible resolvent.**

Were the resolvent reducible, the Galoisian group would not be the smallest group such that rational functions of the x_i remaining under it numerically unaltered have a rational value.

The proposition is true for general equations if we replace the Galoisian group by the symmetric group. That the resolvent then is irreducible will be proved in §70.

Lagrange's Theorem is in the theory of special equations replaced by the **Theorem of Lagrange-Galois,**[1] which we formulate referring rationality to the domain Ω of the equation:

(116) **If a rational function φ_1 in the roots x_i of a special equation remains numerically unaltered by all those permutations of the Galoisian group for the equation which leave another rational function ψ_1 numerically unaltered, then the function φ_1 is rationally expressible in terms of the function ψ_1.**

In other words:

(117) **Any number ω_1 in the domain**

$$\Omega(x_i)_1^n$$

remaining unaltered under a subgroup H of the Galoisian group G is rational in

$$\Omega(\psi_1)$$

if ψ_1 is a function belonging to H.

This follows from Lagrange's Theorem; or we prove it anew setting

$$G = H + Ht_2 + \ldots + Ht_j$$

[1] The Theorem of Lagrange-Galois is also called the Theorem of Lagrange generalized.

and assuming that

$$\psi_1, \psi_2, \ldots, \psi_j$$
$$\omega_1, \omega_2, \ldots, \omega_j$$

are corresponding numbers. The ω_i may or may not be distinct and it is

$$\psi_1 = \psi_1(x_i)$$
$$\omega_1 = \omega_1(x_i).$$

As we did before in §62, we construct the integral function

$$\varphi(\psi)\left(\frac{\omega_1}{\psi - \psi_1} + \frac{\omega_2}{\psi - \psi_2} + \cdots + \frac{\omega_j}{\psi - \psi_j}\right) = \chi(\psi),$$

where

$$\varphi(\psi) = (\psi - \psi_1)(\psi - \psi_2) \ldots (\psi - \psi_j)$$

is rational. Its coefficients are rational in the x_i and unaltered under the Galoisian group which cannot do more than permute the ω_i and ψ_i and permute them alike. Hence the coefficients are rational numbers by property (2) of the Galoisian group, and setting

$$\psi = \psi_1$$

we find as before

$$\omega_1 = R(\psi_1).$$

This completes the adjustments which convert Lagrange's theory of general equations into Galois' theory of special equations, and recalling now the substance of Lagrange's plan we conclude for the plan of Galois:

> **Whenever the Galoisian group of a special equation has a series of subgroups each normal and of prime index in the preceding, the series beginning with the Galoisian group and ending with identity, then we can solve the general equation by algebraic operations using primitive roots of unity.**

But when these conditions do not hold, it would not seem possible to solve the special equation by algebraic operations which beside the rational operations include the extraction of roots.[1]

[1] Compare §§40 and 76. The final statement is in §82.

§70. GENERAL EQUATION

If the complete Galoisian function $G(v)$ of an equation in Ω is reducible there, the equation is said to be **affected** in Ω. We proceed to prove that

(118) **the general equation is unaffected in its domain** composed of numbers that are rational functions of its coefficients

$$c_1, c_2, \ldots, c_n$$

as independent variables.

To this purpose, let first

$$x_1, x_2, \ldots, x_n$$

be independent variables and

$$v_1 = \alpha_1 x_1 + \alpha_2 x_2 + \ldots + \alpha_n x_n$$

a function of these variables with distinct and rational coefficients taking under the symmetric group on the x_i the values

$$v_1, v_2, \ldots, v_{n!}$$

Since the function

$$G(v) = (v - v_1)(v - v_2) \ldots (v - v_{n!})$$

is symmetric in the x_i, it is rational in their elementary symmetric functions which we now replace by the independent variables c_i, and as rational function of v and the c_i we may denote it by

$$G(v|c_i) = (v - v_1)(v - v_2) \ldots (v - v_{n!}).$$

Suppose this function is reducible in the domain of rational numbers, so that

$$G(v|c_i) = g(v|c_i) \cdot g'(v|c_i) \ldots$$

As the function identically disappears for

$$v = v_1$$

when we write v_1 and the c_i in terms of the x_i, a factor

$$g(v|c_i)$$

must identically disappear when we do so. But with

$$g(v_1|c_i) = 0.$$

also

$$g(v_2|c_i) = 0$$

$$\ldots\ldots\ldots$$

$$g(v_{n!}|c_i) = 0,$$

because permutations on the x_i alter v_1 without changing the c_i,[1]

[1] We can apply such permutations inasmuch as $g(v|c_i) = 0$ is an identity in the x_i.

and $g(v|c_i)$ appears to be the same function as $G(v|c_i)$ since it has the same roots.

From this the proposition follows, because $G(v)$ is irreducible in the domain of the c_i when $G(v|c_i)$ is irreducible in the domain of rational numbers.

The Galoisian group of the general equation is the symmetric group. Therefore we infer from proposition (115) that

(119) **the conjugate values which a rational function takes under the symmetric group are roots of a rational and irreducible resolvent of the general equation.**

The Galoisian group of any equation without affect is the symmetric group, and such an equation is in the Galoisian theory sometimes spoken of as a general equation.

An example of an unaffected equation other than the general is

$$x^3 - 2 = 0.$$

It has the roots

$$x_1 = 2^{1/3},\ x_2 = \omega 2^{1/3},\ x_3 = \omega^2 2^{1/3},$$

and its group is symmetric by proposition (109) since every symmetric function of its roots has a rational value while an alternating function like

$$\sqrt{\Delta} = (x_1 - x_2)(x_1 - x_3)(x_2 - x_3) = 6\sqrt{-3}$$

has not.[1] Indeed we find that any rational function of its roots which has a rational value, like

$$x_1x_2 - x_3^2 = 0,$$

remains numerically unaltered under

$$S = 1,\ (12),\ (13),\ (23),\ (123),\ (132).$$

For the general equation the only rational functions of its roots which have a rational value are the symmetric functions.[2] But this is by no means true for all unaffected equations, as appears from the example.

Hence it is inaccurate to say that the plan of Lagrange breaks down whenever functions occur with numerically equal conjugate values, because this cannot happen. This would imply the

[1] Compare the explanation in §73, example (a).

[2] By the definition of its domain following proposition (118); or by proposition (119) since an assymetric function is a root of an irreducible resolvent.

existence of assymetric functions having a rational value,[1] which is not possible with the general equation. But it is true that the plan of Lagrange cannot be applied to all unaffected equations.

§71. DUALITY OF PLANS

It is clear that the theory of Galois is precisely dual to the theory of Lagrange: in the general theory of special equations the Galoisian group takes the place which the symmetric group has in the special theory of general equations, and from there on both theories run precisely parallel.

We put together dually their principal theorems referring rationality to the domain of the equation:

Lagrange	**Galois**
The group of a general equation in x is the symmetric group on its roots x_i.	The group of a special equation in x is the Galoisian group on its roots x_i.
Permutations of the symmetric group leaving a rational function ψ_1 of the x_i formally unaltered compose a subgroup H of the symmetric group.	Permutations of the Galoisian group leaving a rational function ψ_1 of the x_i numerically unaltered compose a subgroup H of the Galoisian group.
To every subgroup H of the symmetric group belongs a rational function ψ_1 remaining under H formally unaltered.	To every subgroup H of the Galoisian group belongs a rational function ψ_1 remaining under H numerically unaltered.
If a group has a subgroup H of index j, a rational function ψ_1 belonging to H takes j formally different conjugate values $\psi_1, \psi_2, \ldots, \psi_j$ under the group which belong to subgroups conjugate with H in the group.	If a group has a subgroup H of index j, a rational function ψ_1 belonging to H takes j numerically different conjugate values $\psi_1, \psi_2, \ldots, \psi_j$ under the group which belong to subgroups conjugate with H in the group.

[1] For instance, the sum of the conjugate values which are distinct.

Lagrange	Galois
If a group has a subgroup of index j, the conjugate values which a rational function belonging to H takes under the group are roots of a resolvent $r(\psi) = 0$, degree j and coefficients unaltered by the group.	If a group has a subgroup of index j, the conjugate values which a rational function belonging to H takes under the group are roots of a resolvent $r(\psi) = 0$, degree j and coefficients unaltered by the group.
The conjugate values which a rational function takes under the symmetric group are roots of a rational and irreducible resolvent.	The conjugate values which a rational function takes under the Galoisian group are roots of a rational and irreducible resolvent.
Theorem of Lagrange: If a rational function φ_1 remains formally unaltered by all those permutations of the symmetric group that leave another rational function ψ_1 formally unaltered, then φ_1 is rationally expressible in terms of ψ_1: $\varphi_1 = R(\psi_1)$.	Theorem of Lagrange-Galois: If a rational function φ_1 remains numerically unaltered by all those permutations of the Galoisian group that leave another rational function ψ_1 numerically unaltered, then φ_1 is rationally expressible in terms of ψ_1: $\varphi_1 = R(\psi_1)$.
If the symmetric group of a general equation is soluble, the equation is solvable by algebraic operations using primitive roots of unity.	If the Galoisian group of a special equation is soluble, the equation is solvable by algebraic operations using primitive roots of unity.

§72. IRREDUCIBLE EQUATION

From the fundamental properties of the Galoisian group of an equation follows that

(120) **the group of an equation $F(x) = 0$ is transitive on those, and only those, roots of the equation which are roots of an irreducible factor $f(x)$ of $F(x)$.**

For is the group intransitive on the roots x_i of the function

$$F(x) = (x - x_1) \; . \; . \; . \; (x - x_k) \; . \; . \; . \; (x - x_m)$$

in Ω connecting x_1 only with

$$x_1, \; . \; . \; . \; , x_k$$

but no other x_i, the coefficients of the function

$$f(x) = (x - x_1) \; . \; . \; . \; (x - x_k)$$

remain numerically unaltered under the group and are numbers in Ω by its property (2). Hence the function $F(x)$ can be reduced in Ω.

Conversely, is the function $F(x)$ reducible in Ω having there the irreducible factor $f(x)$, the coefficients of $f(x)$ are numbers in Ω and remain unaltered under the group by its property (1). Hence the group can be transitive on the roots of $f(x)$ alone.

If the equation

$$F(x) = 0$$

in Ω is reducible there and $F(x)$ has the irreducible factors

$$\begin{aligned} f_\alpha(x) &= (x - \alpha_1) \; . \; . \; . \; (x - \alpha_k) \\ f_\beta(x) &= (x - \beta_1) \; . \; . \; . \; (x - \beta_l) \\ &. \; . \; . \; . \; . \; . \; . \end{aligned}$$

in Ω, so that

$$F(x) = f_\alpha(x) \, . \, f_\beta(x) \; . \; . \; . \; ,$$

we conclude that the group of the equation is intransitive with the intransitive systems

$$\begin{gathered} \alpha_1, \; . \; . \; . \; , \alpha_k \\ \beta_1, \; . \; . \; . \; , \beta_l \\ . \; . \; . \; . \; . \; . \end{gathered}$$

And if the group of the equation is intransitive with these systems, we conclude that the function is reducible as shown.

Hence we can note:

(121) **If an equation $F(x) = 0$ is reducible, its group G is intransitive so that to every irreducible factor of $F(x)$ corresponds an intransitive system of G.**

And conversely:

(122) **If the group G of an equation $F(x) = 0$ is intransitive, the equation is reducible so that to every intransitive system of G corresponds an irreducible factor of $F(x)$.**

Further we can note:

(123) **The group of an equation is transitive when, and only when, the equation is irreducible.**

If the group G of an equation is regular of degree and order n, also the degree of the equation is n. The n roots of the equation then are conjugate functions under G, and each root belongs to a subgroup of index n in G. This is to say that the order of the subgroup is one and the subgroup itself identity.

Since such an equation is irreducible by proposition (123) and its roots are by the Theorem of Lagrange-Galois rational functions of each other, we infer that an equation with a regular group is normal.

This condition on the group is not only necessary but also sufficient, for the roots of an equation must belong to identity to be rational functions of each other.[1] Hence

(124) **the group of a normal equation, and of such an equation alone, is regular.**

§73. APPLICATIONS

To find the Galoisian group of a given equation is extremely difficult, but the propositions that we just proved are helpful. We know that equations may be rational in different domains, and it is clear that for different domains we may obtain different groups.

Examples:

(a) Find the group G of the equation[2]

$$x^3 - 2 = 0$$

for the domain

$$\Omega = (1).$$

Since the equation is irreducible, the group G is transitive and its order is a divisor of $n! = 6$ while a multiple of $n = 3$. This is to say that the order of G is 6 or 3 and G the symmetric or alternating group.

The discriminant of the equation is

$$\Delta = -27c_3^2 = -108,$$

[1] Compare §66.

[2] Compare §70.

and

$$\sqrt{\Delta} = 3c_3\sqrt{-3} = 6\sqrt{-3}.$$

Were G the alternating group, the rational function

$$\sqrt{\Delta} = (x_1 - x_2)(x_1 - x_3)(x_2 - x_3)$$

would have a rational value remaining unaltered under G. This is not true, whence G is the symmetric group:

$$G = 1, (12), (13), (23), (123), (132).$$

(b) Find the group G of the equation

$$x^3 - 2 = 0$$

for the domain

$$\Omega = (\omega),$$

where ω is a primitive cube root of unity.

Now

$$\sqrt{\Delta} = 6\sqrt{-3} = 6(\omega - \omega^2)$$

has a rational value and must remain unaltered under G. Hence G contains no transposition and is the alternating group:

$$G = 1, (123), (132).$$

We observe that G is regular and consequently the equation normal:

$$x_i = R(x_1),$$

which we verify remembering the roots given in §70.

(c) Find the group G of the equation

$$x^4 - 2 = 0$$

for the domain

$$\Omega = (1).$$

Its roots are

$$x_1 = 2^{1/4},\ x_2 = i2^{1/4},\ x_3 = -2^{1/4},\ x_4 = -i2^{1/4}.$$

The rational function

$$x_1x_3 + x_2x_4 = 0$$

has a rational value and must remain unaltered under G; hence we have

$$G = 1, (13), (24), (13)(24), (12)(34), (14)(23), (1234), (1432)$$

or a subgroup.

This means that the order of G is a divisor of 8. Since the equation is irreducible, the order of G is also a multiple of 4;

therefore it is 8 or 4. Were the order 4, then G would be regular and the equation normal. But

$$x_2 \neq R(x_1),$$

whence G stands as given above.

(d) Find the group G of the equation

$$x^4 - 2 = 0$$

for the domain

$$\Omega = (i).$$

Now G is regular of order 4 since the equation is irreducible and

$$x_i = R(x_1).$$

The rational functions

$$\frac{x_1}{x_4} = \frac{x_4}{x_3} = \frac{x_3}{x_2} = \frac{x_2}{x_1} = i$$

have a rational value and must remain unaltered under G. Hence we have

$$G = 1, (1234), (13)(24), (1432).$$

(e) Find the group G of the equation

$$x^4 + 6x^2 + 1 = 0$$

for the domain

$$\Omega = (1).$$

Since the equation is irreducible, the group G is transitive and its order a multiple of 4.

The equation is quadratic in

$$x^2 = y.$$

Therefore we have

$$y_1 y_2 = 1;$$

whence setting

$$x_1 = \sqrt{y_1},\ x_3 = -\sqrt{y_1}$$
$$x_2 = \sqrt{y_2},\ x_4 = -\sqrt{y_2}$$

we find

$$x_2 = \frac{1}{x_1},\ x_3 = -x_1,\ x_4 = -\frac{1}{x_1}.$$

These relations remain true under

$$G = 1, (12)(34), (13)(24), (14)(23).$$

We observe that G is regular and consequently the equation normal, which is verified by the relations between the roots.

(f) Find the group G of the equation

$$x^4 + 6x^2 + 1 = 0$$

for the domain

$$\Omega = (\sqrt{2}).$$

Now the equation written in the form

$$(x^2 + 3)^2 - 8 = 0$$

is reducible:

$$(x^2 + 3 + 2\sqrt{2})(x^2 + 3 - 2\sqrt{2}) = 0,$$

and the group G must be intransitive connecting the roots of the factors only.

From the relations

$$x_2 = \frac{1}{x_1},\ x_3 = -x_1,\ x_4 = -\frac{1}{x_1}$$

follows

$$x_1x_2 = 1,\ x_1x_4 = -1,$$

whence one factor must have the roots x_1 and x_3 and the other factor the roots x_2 and x_4. We set

$$x_1x_3 = 3 \pm 2\sqrt{2},\ x_2x_4 = 3 \mp 2\sqrt{2};$$

the group leaving these rational functions with rational values numerically unaltered and connecting the roots of the factors is

$$G = 1,\ (13)(24).$$

§74. IMPRIMITIVE EQUATION

Suppose that the equation

$$f(x) = 0$$

of degree n in Ω is irreducible there, so that its Galoisian group is transitive by proposition (123), and set

$$x_1 = \alpha_1.$$

If the algebraic domain $\Omega(\alpha_1)$ is imprimitive, it contains an imprimitive number θ_1 equal to some of its conjugate values:

$$\theta_1 = \rho(\alpha_1) = \ .\ .\ .\ = \rho(\alpha_m),$$

and taking under the transitive group of the equation the other conjugate values

$$\theta_2 = \rho(\beta_1) = \ . \ . \ . = \rho(\beta_m)$$
$$. \ . \ . \ . \ . \ . \ . \ .$$

This divides the roots of the equation into the systems

$$\alpha_1, \ . \ . \ . \ , \alpha_m$$
$$\beta_1, \ . \ . \ . \ , \beta_m$$
$$. \ . \ . \ . \ . \ . \ . \ .$$

such that no two roots in different systems are alike.[1]

Since by proposition (107) the relation

$$\rho(\alpha_i) = \rho(\alpha_j)$$

must remain true when operated on by the permutations of the Galoisian group, it follows that those permutations interchange either roots within systems or else entire systems. Hence

(125) **an equation with an imprimitive algebraic domain has an imprimitive group;**

such an equation is called an **imprimitive equation.**[2]

The conjugate values

$$\theta_1, \theta_2, \ . \ . \ . \ , \theta_k$$

are roots of the function

$$\varphi(\theta) = (\theta - \theta_1)(\theta - \theta_2) \ . \ . \ . \ (\theta - \theta_k)$$

which is rational in Ω by property (2) of the Galoisian group, the group not more than permuting the θ_i, and is irreducible there by proposition (115).

Denoting by

$$\omega_1 = S_1(\alpha_i)$$
$$\omega_2 = S_2(\beta_i)$$
$$. \ . \ . \ . \ . \ . \ . \ .$$

symmetric functions on the imprimitive systems, we construct the integral function

$$\varphi(\theta)\left(\frac{\omega_1}{\theta - \theta_1} + \frac{\omega_2}{\theta - \theta_2} + \ . \ . \ . + \frac{\omega_k}{\theta - \theta_k}\right) = \chi(\theta)$$

[1] The β_i are all unlike, with the α_i; and none equals an α_i as from

$$\alpha_i = \beta_j$$

would follow

$$\rho(\alpha_i) = \rho(\beta_j)$$

and two entire systems were indentical.

[2] It can be proved that an equation with a primitive algebraic domain has a primitive group.

whose coefficients, rational in Ω and unaltered under the Galoisian group which cannot do more than permute the ω_i and θ_i alike, are numbers in Ω by property (2) of the group. Setting

$$\theta = \theta_1,$$

we find

$$\omega_1 = R(\theta_1),$$

this relation being rational in Ω.

It follows that the function[1]

$$f_\alpha(x \mid \theta_1) = (x - \alpha_1) \ . \ . \ . \ (x - \alpha_m)$$

is rational in Ω; it is also irreducible there,[2] since those permutations of the Galoisian group which leave θ_1 unaltered are transitive on the α_i. This means that the function

$$f_\alpha(x) = (x - \alpha_1) \ . \ . \ . \ (x - \alpha_m)$$

is rational and irreducible in $\Omega(\theta_1)$.

As the degree of an irreducible equation is also the degree of its algebraic domain, we have the proposition:

(126) **An imprimitive domain $\Omega(\alpha_1)$ of degree n on Ω is identical with the domain $\Omega'(\alpha_1)$ of degree m on Ω if $\Omega' = \Omega(\theta_1)$ is of degree k on Ω and $k \cdot m = n$.**

For the example (d) of §73 we have:

$$\begin{array}{l} \uparrow \ \Omega(2^{1/4}) = \Omega'(2^{1/4}) \uparrow \\ 4 \ \Omega(2^{1/2}) = \Omega' \quad 2 \downarrow \\ \downarrow \qquad \Omega = (i). \end{array}$$

Here

$$\theta_1 = (2^{1/4})^2 = (-2^{1/4})^2$$
$$\theta_2 = (i2^{1/4})^2 = (-i2^{1/4})^2,$$

whence

$$\alpha_1 = 2^{1/4}, \quad \alpha_2 = -2^{1/4}$$
$$\beta_1 = i2^{1/4}, \quad \beta_2 = -i2^{1/4}$$
$$\theta_1 = 2^{1/2}, \quad \theta_2 = -2^{1/2}.$$

§75. REDUCTION OF GROUP

Any equation which is rational in a domain Ω is with stronger reason rational in a domain on Ω obtained by adjunction. The

[1] It is a function also of θ_1.

[2] By proposition (123).

group of the equation may be different for the new domain, and we prove the proposition:

(127) **The group of an equation is reduced to a subgroup when we adjoin to the domain of the equation a rational function of its roots which belongs to the subgroup.**

Let the group G of the equation

$$F(x) = a_0x^n + a_1x^{n-1} + \ . \ . \ . + a_n = 0$$

rational in the domain

$$\Omega = (a_i)$$

of its coefficients, or any wider domain Ω, have a subgroup H and let

$$\psi_1 = \psi_1(x_i)$$

be a function rational in Ω which belongs to H and adjoined to Ω produces the domain

$$\Omega' = (a_i, \psi_1).$$

Since any function χ of the x_i rational in Ω' is a function of ψ_1 and the x_i rational in Ω—which we may express by setting

$$\chi(x_i) = \chi(x_i | \psi_1),$$

it also is a function of the x_i rational in Ω because ψ_1 is.

Assuming first that such a function χ has a value ρ rational in

$$\Omega' = \Omega(\psi_1),$$

so that

$$\rho = \rho(\psi_1)$$

is a function[1] rational in Ω, we have a relation

$$\chi(x_i | \psi_1) = \rho(\psi_1)$$

between the x_i which is rational in Ω. This relation remains true under G by proposition (107), while the function $\rho(\psi_1)$ is unaltered by the subgroup H of G. Hence we infer that also the function $\chi(x_i)$ is unaltered by H.

Assuming second that the function χ remains unaltered by the subgroup H of ψ_1, we infer from the Theorem of Lagrange-Galois that it has a value rational in Ω', whence it appears that H possesses the fundamental properties of the Galoisian group in Ω', and the proposition follows.

[1] Cf. proposition (91).

It implies that

(128) **a normal domain** $\Omega(x_i)_1^n$ **of degree** r_g **on** Ω **is identical with the domain** $\Omega'(x_i)_1^n$ **of degree** r_h **on** Ω **when** $\Omega' = \Omega(\psi_1)$ **is of degree** j **on** Ω **and** $j = r_g/r_h$,

in the notation of §17. For the group of an equation in its normal domain is identity. If its group in Ω^* is H^*, the degree of the resolvent in Ω^* for identity equals the index of identity in H^*, which is the order of H^*, and the degree of this resolvent is also the degree of the normal domain $\Omega^*(x_i)$ on Ω^*.

It is clear that to solve the equation

$$F(x) = 0$$

means to widen its domain by adjunction so that its roots x_i are rationally known and its group G is reduced to identity, therefore means to obtain its normal domain

$$\Omega(v_1) = \Omega(x_i).$$

But we know now that it is not necessary to adjoin at once an irrationality which is primitive there: we may widen the domain Ω of the equation by successive adjunctions reducing its group G from subgroup to subgroup.

Let in the domain Ω of the equation its group be

$$G = 1, \ldots, t$$

and its resolvent

$$g(v) = (v - v_1) \ldots (v - v_t) = 0,$$

let in the domain Ω' its group be

$$H = 1, \ldots, s$$

and its resolvent

$$h(v) = (v - v_1) \ldots (v - v_s) = 0.$$

The function $h(v)$ is rational in Ω' but may be expressed as

$$h(v|\psi_1) = (v - v_1) \ldots (v - v_s)$$

rational in Ω. Since the permutations of H are contained in G, $h(v)$ is a factor of $g(v)$ and $g(v)$ appears to be reducible in Ω'.

Under the group

$$G = H + \ldots + Ht$$

of the equation the function ψ_1 takes j conjugate values

$$\psi_1, \ldots, \psi_t;$$

and the function $h(v|\psi_1)$ goes into

$$h(v|\psi_1), \ldots, h(v|\psi_t)$$

such that

$$h(v|\psi_1) = (v - v_1) \ldots (v - v_s)$$

$$\cdot \quad \cdot \quad \cdot \quad \cdot \quad \cdot \quad \cdot \quad \cdot$$

$$h(v|\psi_t) = (v - v_t) \ldots (v - v_{st})$$

and the unlike[1] sets

$$v_1, \ldots, v_s$$

$$\cdot \quad \cdot \quad \cdot \quad \cdot \quad \cdot \quad \cdot$$

$$v_t, \ldots, v_{st}$$

contain all values of v_1 conjugate under G. Hence we can say that

(129) adjunction reducing the Galoisian group splits the Galoisian resolvent into conjugate factors:

$$\boxed{g(v) = h(v|\psi_1) \ldots h(v|\psi_t)};$$

these factors are rational in

$$\Omega(\psi_1), \ldots, \Omega(\psi_t)$$

and of the same degree r_h in v.

If the subgroup H is normal in G, the domain $\Omega(\psi_1)$ is normal on Ω since we have

$$\psi_t = R(\psi_1)$$

for any ψ_t by the Theorem of Lagrange-Galois; and the Galoisian resolvent under G splits by adjunction of ψ_1 into rational factors each yielding the group H.

If H is not normal in G, a normal domain on Ω is produced[2] when we adjoin the function

$$\theta_1 = \alpha_1\psi_1 + \ldots + \alpha_t\psi_t,$$

the α_i chosen such that θ_1 is altered by every permutation between the ψ_i. The adjunction of θ_1 reduces G by proposition (127) to the greatest common subgroup D of

$$H, \ldots, H_t$$

and is equivalent to the adjunction of all ψ_i by proposition (102).

We have come to another point of vantage from which we can survey the plan of Galois. The essence of solving an equation

[1] Since v_1 takes different values under different permutations of G.

[2] By propositon (101).

is this: by adjoining proper irrationalities we reduce the Galoisian group of the equation and split its Galoisian resolvent.

With this in mind, we turn to resolvents which solved give proper irrationalities. But in doing so we call attention to the circumstance that it will be important not to confuse the Galoisian resolvent $g(v) = 0$ and the resolvent $r(\psi) = 0$ of Lagrange's conception, the ordinary resolvent so to say. The Galoisian resolvent illumines the solution, but the ordinary resolvent is the one that gives it.

The function ψ_1 is a root of the resolvent equation $r(\psi) = 0$ and

(130) **the group of the resolvent equation**

$$\boxed{r(\psi) = 0}$$

is in the domain of the given equation

$$\boxed{F(x) = 0}$$

the group Γ of permutations which the ψ_i undergo when the x_i are operated on by the permutations of G.

For any function of the ψ_i rational in the domain Ω of the given equation is also a function rational there of the x_i:

$$\Phi(\psi_i) = \varphi(x_i),$$

and if Φ remains unaltered under Γ, then φ remains unaltered under G, and conversely.

Therefore Φ remains unaltered under Γ if it has a value rational in Ω, for φ then remains unaltered under G by property (1) of G; and Φ has a value rational in Ω if it remains unaltered under Γ, for φ then has such a value by property (2) of G. Hence the proposition follows because Γ has the fundamental properties of the group in Ω.

We note that the group Γ of the resolvent is transitive since the resolvent is irreducible.[1]

The group Γ is abstractly defined by the factor-group

$$G/H = \Gamma$$

when H is normal in G, else by the factor-group

$$G/D = \Gamma.$$

[1] By propositions (115) and (123).

Consequently the resolvent in θ has the same group as the resolvent in ψ; it is alike in solution although of higher degree when H is not normal.

If the ψ_i belong to subgroups of G having no permutation but identity in common, then

$$D = 1,$$

abstractly

$$\Gamma = G,$$

and the resolvent

$$r(\psi) = 0$$

is called a **total resolvent.** The adjunction of its roots reduces the Galoisian group to identity and splits the Galoisian resolvent into linear factors. This implies the solution of the given equation, yet nothing is gained because the group of a total resolvent is identical with that of the given equation and its solution equally difficult.

We observe that a total resolvent is Galoisian in the wider sense of this term when the ψ_i belong to identity.

If the groups of the ψ_i have more permutations than identity in common, then the resolvent

$$r(\psi) = 0$$

is called a **partial resolvent.** The adjunction of its roots reduces the Galoisian group to a normal subgroup D and splits the Galoisian resolvent into conjugate factors.

Suppose the group G of the given equation has the composition-series

$$G \quad N \quad J \quad . \; . \; . \quad 1.$$

The smallest groups for partial resolvents are defined by the abstract quotients

$$G/N, N/J, \ldots$$

of the series, and it is the purpose of Galois to replace the solution of the given equation by a successive solution of partial resolvents with such groups. Their roots adjoined to the domain of the given equation reduce its group along the series of composition.

Such partial resolvents we may call **resolvents of composition,** and we infer from proposition (124) that

(131) **for soluble groups resolvents of composition are normal.**

For the group Γ of any such resolvent is circular, since it is of prime degree and order:[1]

$$\Gamma = \sigma, \sigma^2, \ldots, \sigma^p, \qquad [\sigma^p = 1$$

where

$$\sigma = (\psi_1 \psi_2 \ldots \psi_p)$$

and p is a factor of composition.

For soluble groups resolvents of composition are even binomial, we now recall,[2] and can be solved by algebraic operations on rational numbers and primitive roots of unity if constructed in the form

$$r(\epsilon, \psi) = 0,$$

where (ϵ, ψ) is Lagrange's solvent.

§76. NATURAL IRRATIONALITY

To solve an equation

$$F(x) = a_0 x^n + a_1 x^{n-1} + \ldots + a_n = 0$$

rational in the domain

$$\Omega = (a_i)$$

of its coefficients, or any larger domain Ω, we must widen its domain to the normal domain

$$\Omega(v_1) = \Omega(x_i)$$

where its roots are rational, its Galoisian group reduced to identity, its Galoisian resolvent split into linear factors.

We achieve this by successive adjunction of numbers contained in the normal domain; such numbers are roots of resolvent equations and called **natural irrationalities** after Kronecker.

Thus we know exactly what we need to solve the equation. Can we solve it? We recall that a sufficient condition is that the group of the equation be soluble. Is it necessary? When this condition holds we can construct and solve resolvents, otherwise we can not. And with a last glimpse of hope we now ask the question: When natural irrationalities cannot be computed, are there perhaps other irrationalities which can solve the equation?

[1] By proposition (25), and a circular group of same degree as the equation is regular by proposition (59).

[2] Proposition (55).

A negative answer is given with the proposition that

(132) **every possible reduction of the Galoisian group is effected by adjunction of natural irrationalities.**

Suppose that adjoining an irrational number y we produce a domain

$$\Omega_y = (a_i, y)$$

in which the Galoisian resolvent is

$$h(v|y) = (v - v_1) \ldots (v - v_s) = 0.$$

We proved for any domain that the permutations yielded by the resolvent compose a unique group

$$H = 1, \ldots, s.$$

If now $h(v|y)$ is a factor of $g(v)$, then H is a subgroup of G and we reduce the Galoisian group by adjoining y. But the same reduction, of course, results when we adjoin any function ψ of the x_i belonging to H. And furthermore the domain

$$\Omega_\psi = (a_i, \psi)$$

is contained in the domain

$$\Omega_y = (a_i, y),$$

since the coefficients of $h(v|y)$ are rational in no smaller domain on Ω than Ω_ψ and also are rational in Ω_y.[1] It appears that we can certainly construct ψ if we can construct y, and that no solution of the equation is possible which cannot be effected also by natural irrationalities.

Considering that any complete solution of the equation reduces its group to identity, we now conclude for all equations, since the theory of Galois includes that of Lagrange:

> **Whenever the Galoisian group of an equation has a series of subgroups each normal and of prime index in the preceding, the series beginning with the Galoisian group and ending with identity, then we can solve the equation by algebraic operations on rational numbers and primitive roots of unity. But when these conditions do not hold, then it is not possible to solve the equation by algebraic operations on such numbers.**[2]

[1] For the coefficients belong to H, and so does ψ; moreover ψ is a root of an irreducible resolvent and primitive in Ω_ψ.

[2] Compare §§40 and 69. The final statement is in §82. For primitive roots of unity compare footnote to §40.

CHAPTER XII

SPECIAL EQUATIONS

§77. ABELIAN EQUATION

As we did in the preceding chapter, we shall assume here that the equations we treat of have no double roots. Furthermore it will be observed that the typical equations we treat of are irreducible.

An irreducible equation whose group is Abelian we call an **Abelian equation.**

Suppose that

$$f(x) = 0$$

is an Abelian equation with n roots x_i. Then its group G is transitive by proposition (123) and contains a permutation s_i converting x_1 into x_i.

Those permutations of G that leave x_1 unaltered compose a subgroup H, and those that leave x_i unaltered compose a subgroup

$$H_i = s_i^{-1}Hs_i$$

conjugate with H. But subgroups of an Abelian group are normal, by proposition (32),[1] and we have

$$H_i = H = 1,$$

for identity alone leaves every x_i unaltered.

If a permutation s_i' other than s_i would convert x_1 into x_i, then the permutation $s_i's_i^{-1}$ would leave x_1 unaltered, so that

$$s_i's_i^{-1} = 1$$

and

$$s_i' = s_i.$$

Hence the group of the equation is

$$G = 1, s_2, \ . \ . \ . \ , s_n.$$

The group is regular, and it follows by proposition (124) that

(133) **every Abelian equation is normal.**

[1] Cf. also §56.

This means that all roots x_i are rational functions of x_1, and we can set

$$x_1 = R_1(x_1),\ x_2 = R_2(x_1),\ .\ .\ .\ ,\ x_n = R_n(x_1).$$

By proposition (107), any relations

$$x_i = R_i(x_1),\ x_k = R_k(x_1)$$

remain true when we apply the permutation s_k of the group to one and the permutation s_i of the group to the other:

$$x_{ik} = R_i(x_k),\ x_{ki} = R_k(x_i).$$

But with

$$s_i s_k = s_k s_i,$$

which is true for an Abelian group, also

$$x_{ik} = x_{ki}$$

and

$$R_i(x_k) = R_k(x_i).$$

Hence

(134) **for an Abelian equation in x the relation holds:**

$$\boxed{R_i[R_k(x_1)] = R_k[R_i(x_1)]}.$$

Conversely,

(135) **when this relation holds for the roots x_i of an equation while any**

$$\boxed{x_i = R_i(x_1)},$$

the group of the equation is Abelian.

For it follows from

$$x_{ik} = R_i(x_k),\ x_{ki} = R_k(x_i)$$

that

$$x_{ik} = x_{ki}$$

and

$$s_i s_k = s_k s_i,$$

which proves the proposition.

It is to be noted that in proving the converse we did not qualify the equation as irreducible.

§78. CYCLIC EQUATION

An irreducible equation whose group is cyclic we call a **cyclic equation.**

Suppose that

$$f(x) = 0$$

is a cyclic equation with n roots x_i. Its group is cyclic by definition and transitive by proposition (123), therefore by proposition (59) circular:

$$G = \{s\},$$

containing all powers of the circular permutation

$$s = (x_1x_2 \ . \ . \ . \ x_n).$$

It follows from propositions (25) and (124) that

(136) **every normal equation of prime degree is cyclic;**
in particular resolvents of composition for soluble groups are cyclic equations by proposition (131).

Since cyclic groups are Abelian, cyclic equations are also; consequently cyclic equations are normal by proposition (133).

Whether the group of the equation

$$f(x) = 0$$

is G or a subgroup

$$H = \{s^k\}$$

of G, the integral function

$$f(x)\left(\frac{x_2}{x - x_1} + \frac{x_3}{x - x_2} + \ . \ . \ . \ + \frac{x_1}{x - x_n}\right) = \chi(x)$$

has coefficients unaltered by the group and rational by its property (2). Setting

$$x = x_i$$

and

$$\frac{\chi(x_i)}{f'(x_i)} = R(x_i),$$

we find[1] that

$$x_2 = R(x_1),\ x_3 = R(x_2),\ . \ . \ . \ ,\ x_n = R(x_{n-1}),\ x_1 = R(x_n).$$

If we put

$$R[R(x_1)] = R^2(x_1),\ R[R^2(x_1)] = R^3(x_1),\ . \ . \ .$$

where exponents do not denote powers, we have the relations

$$x_2 = R(x_1),\ x_3 = R^2(x_1),\ . \ . \ . \ ,\ x_n = R^{n-1}(x_1),\ x_1 = R^n(x_1),$$

or

$$x_{z+1} = R^z(x_1)$$

with

$$z \equiv 1, 2, \ . \ . \ . \ (\text{mod } n),$$

[1] Compare the procedure in §62.

and say that

(137) **the roots of an equation compose a cycle when the group of the equation is circular of same degree as the equation or a subgroup of the circular group.**

In case of a cyclic equation we obtain the cyclic relation of its roots more simply by applying the permutations of G to the relation

$$x_2 = R(x_1),$$

which is true for any normal equation.

Conversely,

(138) **when the roots of an equation compose a cycle, the group of the equation is circular of same degree as the equation or a subgroup of the circular group.**

For let any permutation of the group be

$$t = \begin{pmatrix} x_1 & \cdots & x_{z+1} & \cdots \\ x_\alpha & \cdots & x_{\zeta+1} & \cdots \end{pmatrix}.$$

Applied to the cyclic relation[1] it gives

$$x_{\zeta+1} = R^z(x_\alpha)$$

by proposition (107). But

$$R^z(x_\alpha) = R^z[R^\alpha(x_1)] = R^{z+\alpha}(x_1) = x_{z+\alpha},$$

whence

$$x_{\zeta+1} = x_{z+\alpha}$$

and

$$\zeta + 1 \equiv z + \alpha \pmod{n}.$$

It appears that

$$t = \begin{pmatrix} 1 & 2 & 3 & \cdots \\ \alpha & \alpha+1 & \alpha+2 & \cdots \end{pmatrix}.$$

With the modulus n understood it is

$$t = \begin{pmatrix} z+1 \\ z+\alpha \end{pmatrix};$$

but this is the $(\alpha - 1)$st power of the circular permutation[2]

$$s = \begin{pmatrix} z \\ z+1 \end{pmatrix} = (12 \ldots n).$$

Hence the group of the equation is G or a subgroup H, and the proposition follows. If the group of the equation is G, the equation is cyclic.

[1] $x_{z+1} = R^z(x_1)$.
[2] Cf. §58.

§79. ROOTS OF UNITY

To solve an equation by algebraic operations, we need prime roots of unity which are primitive, as we recall.

Roots of unity are defined by an equation

$$x^n - 1 = 0;$$

there are n distinct roots since the discriminant[1] of this equation cannot vanish. Such roots whose n-th powers, and no lower ones, equal unity are called **primitive n-th roots of unity.**

If ϵ is some root of this equation, its powers are also, for

$$(\epsilon^\alpha)^n = (\epsilon^n)^\alpha = 1.$$

And if m is its lowest power that equals unity, we say that ϵ belongs to the exponent m and find that

$$\epsilon, \epsilon^2, \ldots, \epsilon^m$$

are all unlike. For were

$$\epsilon^\alpha = \epsilon^\beta,$$

such that

$$\alpha > \beta,$$

from

$$\epsilon^{\alpha-\beta} = 1$$

would follow that a power of ϵ lower than m is unity.

Since

$$\epsilon^{\alpha+m} = \epsilon^\alpha,$$

we can arrange the n roots

$$\begin{array}{llll} \epsilon, & \epsilon^2, & \ldots & , \epsilon^m \\ \epsilon^{m+1}, & \epsilon^{m+2}, & \ldots & , \epsilon^{2m} \\ \cdot & \cdot & \cdot & \cdot \\ \epsilon^{km+1}, & \epsilon^{km+2}, & \ldots & , \epsilon^n \end{array}$$

of the equation so that roots in the same column have the same value. It then appears that m is a divisor of n since

$$(k+1)m = n.$$

Whenever

$$n = p$$

where p is prime, also

$$m = p$$

while

$$k = 0,$$

[1] Cf. §27.

and ϵ is a primitive root of unity whose powers

$$\epsilon, \epsilon^2, \ldots, \epsilon^p$$

are all distinct and identical with

$$\epsilon^\alpha, \epsilon^{2\alpha}, \ldots, \epsilon^{p\alpha}$$

except for the sequence.

Hence it appears that

(139) **a prime root of unity other than 1 is primitive.**

The equation

$$c(x) = \frac{x^p - 1}{x - 1} = x^{p-1} + x^{p-2} + \ldots + x + 1 = 0$$

whose root it is we call a **cyclotomic equation.**

But before we can approach this equation, we must digress into the theory of numbers.

§80. CONGRUENCE

Let $\varphi(n)$ denote the number of positive integers smaller than n which are prime to n.

When

$$n = p$$

is a prime number, then

$$1, 2, \ldots, p - 1$$

all are prime to p and

$$\varphi(p) = p - 1.$$

When

$$n = p^\alpha,$$

we eliminate from

$$0, 1, 2, \ldots, n - 1$$

the n/p numbers

$$0, p, 2p, \ldots, \left(\frac{n}{p} - 1\right)p,$$

leaving those prime to n, and find

$$\varphi(n) = n - \frac{n}{p} = n\left(1 - \frac{1}{p}\right).$$

When

$$n = ab$$

where a and b are one prime to the other, we put

$$z = ay - bx$$

and let

$$x = 0, 1, 2, \ldots, a - 1$$
$$y = 0, 1, 2, \ldots, b - 1.$$

This gives n values for z which are all distinct modulo n. For assuming that

$$z \equiv z'(\text{mod } n),$$

we have[1]

$$z - z' = a(y - y') - b(x - x')$$

divisible by n and consequently $b(x - x')$ by a. But b is prime to a, hence $x - x'$ is divisible by a and we have

$$x - x' = 0,$$

as it is smaller than a. It appears that

$$x = x',$$

and similarly that

$$y = y'.$$

Thus

$$z \equiv 0, 1, 2, \ldots, n - 1 \ (\text{mod } n).$$

To eliminate those values of z that are not prime to n, we must eliminate those values of x that are not prime to a and those values of y that are not prime to b, whence

$$\varphi(n) = \varphi(a)\varphi(b).$$

When

$$a = p^\alpha$$

and

$$b = q^\beta$$

where p and q are prime, then a and b are one prime to the other, and we have the proposition:

(140) **If there are $\varphi(n)$ positive integers smaller than n which are prime to n and**

$$\boxed{n = p^\alpha q^\beta},$$

then

$$\boxed{\varphi(n) = n\left(1 - \frac{1}{p}\right)\left(1 - \frac{1}{q}\right)}.$$

Among the values of z one must be

$$z = c' + kn$$

where c' is an assigned number. Hence the equation

$$ay - bx = c' + kn$$

[1] $z' = ay' - bx'$.

has one solution; and so has any **Diophantine equation**

$$ay - bx = c$$

where a and b are one prime to the other.

It follows that $bx + c$ is divisible by a and the linear congruence

$$bx + c \equiv 0(\text{mod } a)$$

has a solution, and only one. This is to say that there is one value of x satisfying the congruence and at the same time determining y; it is said to be a **root** of the congruence.

Likewise, for any integral function

$$f(x) = a_0x^n + a_1x^{n-1} + \ldots + a_n$$

with integral coefficients and a_0 not divisible by p the congruence

$$f(x) \equiv 0(\text{mod } p)$$

is said to have a root α if $f(\alpha)$ is divisible by p.

Using such a function of degree n, we prove that

(141) **a congruence of degree n modulo p where p is prime has no more than n roots.**

This is true for

$$n = 1,$$

and the proposition follows by complete induction.

Assuming that x and α are indeterminate, we can set[1]

$$f(x) = (x - \alpha)g(x) + f(\alpha),$$

where $g(x)$ has the same qualifications as $f(x)$ but is of degree $n - 1$. From this relation we conclude that $(x - \alpha)g(x)$ is divisible by p if $f(x)$ and $f(\alpha)$ are, so that $g(x)$ then is divisible by p if x is not congruent to α.

If now the proposition is true for $n - 1$, then $g(x)$ has no more than $n - 1$ incongruent roots and consequently $f(x)$ no more than n such roots, whence the proposition is true also for n. The additional root must be α making both terms on the right divisible by p although $g(\alpha)$ is not divisible.

But differing from equations, a congruence may have less than n roots; and even no roots at all like

$$x^2 + 1 \equiv 0(\text{mod } 3).$$

Yet there exists a peculiar congruence that cannot have less than n roots, as we shall see presently.

[1] Cf. §5.

§81. FERMAT'S THEOREM

According to **Fermat's Theorem,**[1]

(142) for a prime number p one has

$$\boxed{a^{p-1} \equiv 1 \pmod{p}}$$

if a is a number not divisible by p.

This is true because it can be proved by complete induction that

$$a^p \equiv a (\text{mod } p).$$

For

$$a = 1$$

evidently

$$1^p \equiv 1 (\text{mod } p).$$

For

$$a = 2$$

we have

$$2^p = (1 + 1)^p = 1 + p + \frac{p(p-1)}{2} + \ldots + p + 1$$
$$\equiv 2 (\text{mod } p)$$

since all middle terms are divisible by p. And if it is true that

$$a^p \equiv a (\text{mod } p),$$

then also

$$(a + 1)^p = a^p + pa^{p-1} + \frac{p(p-1)}{2} a^{p-2} + \ldots + pa + 1$$
$$\equiv a + 1 (\text{mod } p).$$

It appears that the congruence

$$x^p \equiv x (\text{mod } p)$$

has p roots

$$0, 1, 2, \ldots, p - 1,$$

and the congruence

$$x^{p-1} \equiv 1 (\text{mod } p)$$

has $p - 1$ roots

$$1, 2, \ldots, p - 1.$$

The two congruences are essentially alike since the first has only an additional root 0. These are the peculiar congruences with a number of roots equal to their degree.

If m is the smallest positive number for which

$$a^m \equiv 1 (\text{mod } p),$$

[1] Fermat lived 1601–1665.

then a is said to belong to the exponent m modulo p, and if also

$$a^n \equiv 1 (\text{mod } p),$$

then n is a multiple of m, so that m always is a divisor of $p - 1$.

A number g that belongs to the exponent $p - 1$ is called a **primitive root of the prime number p.**

For the prime number $p = 2$ it is readily seen that $g = 1$. If p is an odd prime,[1] we set

$$p - 1 = a^\alpha b^\beta c^\gamma \ldots$$

where a, b, c are prime numbers and prove that

(143) **there is a number A belonging to the exponent a^α.**

For the congruence

$$x^{\frac{p-1}{a}} \equiv 1 (\text{mod } p)$$

is of lower degree than $p - 1$ and has by proposition (141) fewer than $p - 1$ roots. Hence there is a number y among

$$1, 2, \ldots, p - 1$$

which does not satisfy the congruence. We set

$$A = y^{b^\beta c^\gamma \ldots}$$

and have

$$A^{a^\alpha} \equiv 1 (\text{mod } p).$$

If now A belongs to the exponent m, then m is a divisor of a^α and consequently a power of a. It must be equal to a^α, for were

$$m < a^\alpha,$$

we should have[2]

$$A^{a^{\alpha-1}} \equiv 1$$

and

$$y^{a^{\alpha-1} b^\beta c^\gamma \ldots} = y^{\frac{p-1}{a}} \equiv 1 (\text{mod } p),$$

which is a contradiction.

Likewise it can be shown that there are numbers $B, C, \ldots$ belonging to the exponents $b^\beta, c^\gamma, \ldots$ respectively.

We now prove that

(144) **the product $g = A\ B\ C \ldots$ belongs to the exponent $p - 1$.**

It cannot belong to an exponent h such that

$$h < p - 1,$$

[1] An odd prime is any prime number other than 2.

[2] A^m to some power a^k would give $A^{a^{\alpha-1}}$.

for that gives[1]

$$\frac{p-1}{h} = j$$

where j is a positive integer greater than 1 and hence divisible by some prime number $a, b, c, \ldots$ Suppose it is divisible by a; then it follows from

$$g^h \equiv 1$$

that

$$g^{\frac{p-1}{a}} \equiv 1 (\text{mod } p)$$

since $(p-1)/a$ is divisible by h if j is divisible by a. In other words we then have

$$A^{\frac{p-1}{a}} B^{\frac{p-1}{a}} C^{\frac{p-1}{a}} \ldots \equiv 1.$$

But $(p-1)/a$ is divisible by $b^\beta, c^\gamma, \ldots$ Hence

$$B^{\frac{p-1}{a}} \equiv 1, C^{\frac{p-1}{a}} \equiv 1$$

leaving

$$A^{\frac{p-1}{a}} \equiv 1 (\text{mod } p).$$

This is not true since $(p - 1)/a$ is not divisible by a^α, and the proposition follows.[2]

It appears that g is a primitive root of the prime number p. Its powers

$$1, g, g^2, \ldots, g^{p-2}$$

are identical with

$$1, 2, 3, \ldots, p-1$$

except for the sequence, since they are $p - 1$ incongruent numbers not divisible by p.

If $p - 1$ and k have the greatest common factor f and

$$p - 1 = fQ$$
$$k = fq,$$

where

$$q < Q$$

and prime to it, then

$$g^{kl} = g^{fql} \equiv 1$$

[1] $g^{p-1} = (ABC \ldots)^{p-1} \equiv 1$ by proposition (143).

[2] This proof was given by the great Gauss; he lived 1777–1855.

only when l is divisible by Q. Hence g^k belongs to the exponent Q, and as q can have $\varphi(Q)$ values, there are $\varphi(Q)$ incongruent numbers belonging to the exponent Q.

For

$$Q = p - 1$$

we find that

(145) **there are $\varphi(p - 1)$ primitive roots of p.**

§82. CYCLOTOMIC EQUATION

Returning to the cyclotomic equation for prime roots of unity, we prove for odd primes that[1]

(146) **a cyclotomic equation for prime roots of unity is irreducible in the domain of rational numbers.**

In

$$c(x) = \frac{x^p - 1}{x - 1}$$

we set

$$x = y + 1.$$

Then

$$c(y + 1) = y^{p-1} + py^{p-2} + \frac{p(p-1)}{2}y^{p-3} + \dots + \frac{p(p-1)}{2}y + p.$$

The right member is irreducible by proposition (9), and with $c(y + 1)$ also $c(x)$ is irreducible.

Hence the group G of the cyclotomic equation

$$c(x) = x^{p-1} + x^{p-2} + \dots + x + 1 = 0$$

is transitive[2] and its order a multiple[3] of $p - 1$.

The roots of this equation are

$$x_1 = \epsilon,\ x_2 = \epsilon^2,\ \dots,\ x_{p-1} = \epsilon^{p-1}, \qquad [\epsilon^p = 1$$

so that

$$x_i = x_1{}^i,$$

and a permutation of G leaving x_1 unaltered leaves unaltered every x_i. It necessarily is identity, whence G is regular of order[4]

$$r = p - 1.$$

[1] The same can be proved for any cyclotomic equation.

[2] By proposition (123).

[3] By proposition (60).

[4] Cf. §§72 and 77. It follows by proposition (124) that a cyclotomic equation is normal.

By proposition (107) the relation

$$x_i = x_1{}^i$$

must remain true if operated on by the permutations of G. But a permutation of G converting

$$x_1 \rightarrow x_i = x_1{}^i$$

converts

$$x_1{}^i \rightarrow x_i{}^i = x_1{}^{i^2}$$

$$. \quad . \quad . \quad . \quad . \quad . \quad .$$

and is modulo p

$$s = (x_1 x_1{}^i x_1{}^{i^2} \ . \ . \ . \ ) \ . \ . \ .$$

If we now take as i a primitive root of p:

$$i = g,$$

then one cycle of s contains all x_i; it is modulo p

$$s = (\epsilon \epsilon^g \epsilon^{g^2} \ . \ . \ . \ ) = (x_1 x_g x_{g^2} \ . \ . \ . \ )$$

of order $p - 1$, whence

$$G = \{s\}.$$

This means that

(147) **every cyclotomic equation for prime roots of unity is cyclic**

and implies that every cyclotomic equation for prime roots of unity is solvable.[1] Besides it is of degree one less than the degree of the resolvent for which we need its roots.[2]

Hence prime roots of unity can be computed by algebraic operations, and we can state the main result of our study in the final form:[3]

(148) **Whenever the group of an equation has a series of subgroups each normal and of prime index in the preceding, the series beginning with the group and ending with identity, then we can solve the equation by algebraic operations which beside the rational operations include the extraction of roots. But when these conditions do not hold, then it is not possible to solve the equation by algebraic operations on numbers which are rationally known.**

[1] Cf. proposition (76).

[2] Note that for a cubic resolvent the cyclotomic equation is of degree two, that is one less. This is why we can solve the general cubic.

[3] Compare §76, also §§40 and 69.

Thus not every algebraic equation has algebraic roots, and the fundamental theorem of algebra is not a theorem of algebra at all.

§83. DISCRIMINANT OF CYCLOTOMIC EQUATION

The discriminant Δ of the cyclotomic equation

$$c(x) = x^{p-1} + x^{p-2} + \ldots + x + 1 = 0$$

is determined by the relation[1]

$$(-1)^{\frac{p-1}{2}} \Delta = c'(x_1)c'(x_2) \ldots c'(x_{p-1})$$

since

$$(-1)^{\frac{p-1}{2}} = (-1)^{\frac{(p-1)(p-2)}{2}},$$

$p - 1$ being even and $p - 2$ odd.

Differentiating

$$x^p - 1 = (x - 1)c(x)$$

we obtain

$$px^{p-1} = c(x) + (x - 1)c'(x),$$

and setting

$$x = x_i$$

with x_i standing for any root of the cyclotomic equation, we have

$$px_i^{p-1} = (x_i - 1)c'(x_i).$$

Multiplying such relations for all x_i, we find that

$$p^{p-1} = \prod_{i=1}^{p-1} c'(x_i) \prod_{i=1}^{p-1} (1 - x_i) = (-1)^{\frac{p-1}{2}} \Delta \,.\, c(1)$$

since

$$\prod x_i = 1$$

and $p - 1$ is even. But

$$c(1) = p,$$

whence

(149) **the discriminant Δ of a cyclotomic equation for prime *p*-th roots of unity has the value**

$$\boxed{\Delta = (-1)^{\frac{p-1}{2}} p^{p-2}}.$$

It follows that

$$\sqrt{\Delta} = \pm p^{\frac{p-3}{2}} i^{\frac{p-1}{2}} \sqrt{p}$$

[1] Cf. §27.

with rational $p^{(p-3)/2}$ is real if

$$p \equiv 1 \pmod 4$$

and imaginary if

$$p \equiv 3 \pmod 4.$$

§84. APPLICATIONS

It is in the sense of Galois to use cyclic resolvents, but often convenient to solve cyclotomic equations by more elementary methods.

In the computations which follow we shall consider that

$$\epsilon^k = \cos\frac{2k\pi}{p} + i\sin\frac{2k\pi}{p}$$

whenever difficulties arise concerning the sign.

To find primitive roots of p that we shall need, there is no general method but trying.

Examples:

(*a*) For

$$x^4 + x^3 + x^2 + x + 1 = 0$$

and

$$p = 5$$

we can take

$$g \equiv 2 \pmod 5$$

giving

$$g^2 \equiv 4,\ g^3 \equiv 3,\ g^4 \equiv 1.$$

Hence

$$s = (\epsilon\epsilon^2\epsilon^4\epsilon^3) = (1243)$$

and

$$\begin{aligned} G_4 &= 1,\ (1243),\ (14)(23),\ (1342) \\ G_2 &= 1,\ (14)(23) \\ G_1 &= 1. \end{aligned}$$

To G_2 belongs the function

$$\varphi_1 = x_1 + x_4$$

with the conjugate value

$$\varphi_2 = x_2 + x_3.$$

We have

$$\varphi_1 + \varphi_2 = x_1 + x_2 + x_3 + x_4 = -1,$$

and also

$$\varphi_1\varphi_2 = x_1 + x_2 + x_3 + x_4 = -1$$

since

$$x_i x_k = \epsilon^{i+k \pmod 5} = x_{i+k \pmod 5},$$

so that the φ_i are roots of the quadratic resolvent

$$\varphi^2 + \varphi - 1 = 0$$

giving

$$\varphi_1 = \frac{-1 + \sqrt{5}}{2} \qquad [\varphi_1 = 2\cos\frac{2\pi}{5}$$

$$\varphi_2 = \frac{-1 - \sqrt{5}}{2}. \qquad [\varphi_2 = -2\cos\frac{\pi}{5}$$

To G_1 belongs the function

$$\psi_1 = x_1$$

with the conjugate value

$$\psi_2 = x_4.$$

But

$$\psi_1 + \psi_2 = \varphi_1$$

$$\psi_1\psi_2 = \epsilon^0 = 1,$$

so that the ψ_i are roots of the quadratic resolvent

$$\psi^2 + \frac{1 - \sqrt{5}}{2}\psi + 1 = 0$$

giving

$$\psi_1 = x_1 = \frac{-1 + \sqrt{5} + \sqrt{-10 - 2\sqrt{5}}}{4} \qquad [\cos\frac{2\pi}{5} + i\sin\frac{2\pi}{5}$$

$$\psi_2 = x_4 = \frac{-1 + \sqrt{5} - \sqrt{-10 - 2\sqrt{5}}}{4}. \qquad [\cos\frac{2\pi}{5} - i\sin\frac{2\pi}{5}$$

Likewise

$$x_2 = \frac{-1 - \sqrt{5} + \sqrt{-10 + 2\sqrt{5}}}{4} \qquad [-\cos\frac{\pi}{5} + i\sin\frac{\pi}{5}$$

$$x_3 = \frac{-1 - \sqrt{5} - \sqrt{-10 + 2\sqrt{5}}}{4}. \qquad [-\cos\frac{\pi}{5} - i\sin\frac{\pi}{5}$$

(*b*) For

$$x^6 + x^5 + x^4 + x^3 + x^2 + x + 1 = 0$$

and

$$p = 7$$

we can take

$$g \equiv 3 \pmod 7$$

giving

$$g^2 \equiv 2,\ g^3 \equiv 6,\ g^4 \equiv 4,\ g^5 \equiv 5,\ g^6 \equiv 1.$$

Hence

$$s = (132645)$$

and

$G_6 = 1, (132645), (124)(365), (16)(34)(25), (142)(356), (154623)$
$G_3 = 1, (124)(365), (142)(356)$
$G_1 = 1.$

To G_3 belongs the function

$$\varphi_1 = x_1 + x_2 + x_4$$

with the conjugate value

$$\varphi_2 = x_3 + x_6 + x_5$$

and

$$\varphi_1 + \varphi_2 = -1$$
$$\varphi_1\varphi_2 = 2.$$

From the quadratic resolvent

$$\varphi^2 + \varphi + 2 = 0$$

we obtain

$$\varphi_1 = \frac{-1 + \sqrt{-7}}{2} \qquad [\sin\frac{2\pi}{7} + \sin\frac{3\pi}{7} - \sin\frac{\pi}{7} > 0$$

$$\varphi_2 = \frac{-1 - \sqrt{-7}}{2}. \qquad [\sin\frac{\pi}{7} - \sin\frac{2\pi}{7} - \sin\frac{3\pi}{7} < 0$$

To G_1 belongs the function

$$\psi_1 = x_1$$

with the conjugate values

$$\psi_2 = x_2$$
$$\psi_3 = x_4,$$

and these functions are roots of the cubic resolvent

$$\psi^3 - \varphi_1\psi^2 + \varphi_2\psi - 1 = 0$$

which we know how to solve.

(*c*) Historic interest is attached to the cyclotomic equation

$$x^{16} + x^{15} + \ldots + x + 1 = 0$$

with

$$p = 17,$$

because here Gauss discovered that the construction of a regular polygon with rule and compass is possible whenever the roots of unity which give its vertices can be computed by quadratic resolvents.

Taking

$$g \equiv 3 (\text{mod } 17)$$

we obtain

$$s = (1\ 3\ 9\ 10\ 13\ 5\ 15\ 11\ 16\ 14\ 8\ 7\ 4\ 12\ 2\ 6)$$

and

$$G_{16} = \{s\}, G_8 = \{s^2\}, G_4 = \{s^4\}, G_2 = \{s^8\}, G_1 = 1.$$

Setting

$$\varphi_1 = x_1 + x_9 + x_{13} + x_{15} + x_{16} + x_8 + x_4 + x_2$$
$$\varphi_2 = x_3 + x_{10} + x_5 + x_{11} + x_{14} + x_7 + x_{12} + x_6$$

we have

$$\varphi_1 + \varphi_2 = -1$$
$$\varphi_1\varphi_2 = -4.$$

From the quadratic resolvent

$$\varphi^2 + \varphi - 4 = 0$$

we find

$$\varphi_1 = \frac{-1 + \sqrt{17}}{2}$$

$$\varphi_2 = \frac{-1 - \sqrt{17}}{2}.$$

Setting

$$\chi_1 = x_1 + x_{13} + x_{16} + x_4$$
$$\chi_2 = x_9 + x_{15} + x_8 + x_2$$
$$\chi_3 = x_3 + x_5 + x_{14} + x_{12}$$
$$\chi_4 = x_{10} + x_{11} + x_7 + x_6$$

we have

$$\chi_1 + \chi_2 = \varphi_1$$
$$\chi_1\chi_2 = -1$$

and

$$\chi_3 + \chi_4 = \varphi_2$$
$$\chi_3\chi_4 = -1.$$

From the quadratic resolvent

$$\chi^2 + \frac{1 - \sqrt{17}}{2}\chi - 1 = 0$$

we find

$$\chi_1 = \frac{-1 + \sqrt{17}}{4} + \sqrt{\frac{17 - \sqrt{17}}{8}}$$

$$\chi_2 = \frac{-1 + \sqrt{17}}{4} - \sqrt{\frac{17 - \sqrt{17}}{8}},$$

and from the quadratic resolvent

$$\chi^2 + \frac{1+\sqrt{17}}{2}\chi - 1 = 0$$

we find

$$\chi_3 = \frac{-1-\sqrt{17}}{4} + \sqrt{\frac{17+\sqrt{17}}{8}}$$

$$\chi_4 = \frac{-1-\sqrt{17}}{4} - \sqrt{\frac{17+\sqrt{17}}{8}}.$$

Setting

$$\psi_1 = x_1 + x_{16}$$

$$\psi_2 = x_{13} + x_4$$

we have

$$\psi_1 + \psi_2 = \chi_1$$

$$\psi_1\psi_2 = \chi_3,$$

and from the quadratic resolvent

$$\psi^2 - \chi_1\psi + \chi_3 = 0$$

we find ψ_1 and ψ_2. Finally we set

$$\omega_1 = x_1$$

$$\omega_2 = x_{16}$$

where

$$\omega_1 + \omega_2 = \psi_1$$

$$\omega_1\omega_2 = 1,$$

and from the quadratic resolvent

$$\omega^2 - \psi_1\omega + 1 = 0$$

we find x_1 and x_{16}.

Since the roots of a cyclotomic equation can be constructed geometrically with rule and compass whenever they can be computed algebraically by square roots, this might seem possible whenever

$$p - 1 = 2^m$$

and

$$p = 2^m + 1,$$

for a cyclic group of order $p - 1$ has a subgroup of any order which is a divisor of $p - 1$, as we know.[1]

But when

$$m = kq$$

[1] Cf. proposition (75).

where q is odd, then $2^m + 1$ is not prime because

$$\frac{(2^k)^q + 1}{2^k + 1} = (2^k)^{q-1} - (2^k)^{q-2} + \cdots + 1$$

is integral. Hence it is necessary that

$$m = 2^n$$

and

$$p = 2^{2^n} + 1.$$

Such numbers p are called **Fermat numbers,** and it was believed that they all are prime until Euler[1] discovered that $2^{2^5} + 1$ is divisible by 641.[2]

§85. METACYCLIC EQUATION

An equation whose group is metacyclic we call a **metacyclic equation.**[3] We do not have to qualify such an equation as irreducible because a metacyclic group is transitive.[4]

When the degree of a metacyclic equation is prime, the equation is solvable by proposition (86); and so is any equation of prime degree whose group is a transitive subgroup of the metacyclic. Conversely, when an irreducible equation of prime degree is solvable, its group is by proposition (86) the metacyclic group or one of its transitive subgroups.

This gives the **Theorem of Galois:**

(150) **If an irreducible equation of prime degree is solvable, its group is the metacyclic group or a transitive subgroup of the metacyclic group.**

When we reduce the group of a metacyclic equation of prime degree along its series of composition, the equation remains irreducible until the last by proposition (84). Only when we reach identity, the equation splits and splits at once into linear factors yielding the roots. The same is true for any equation of prime degree whose group is a transitive subgroup of the metacyclic, is therefore true for any irreducible equation of prime degree which is solvable.

[1] Euler lived 1707–1783.

[2] Now Fermat numbers are known not to be prime for

$$n = 5, 6, 7, 8, 9, 11, 12, 23.$$

[3] Compare footnotes to §§55 and 58.

[4] Cf. proposition (123).

Such equations have a peculiar property:

(151) **All roots of an irreducible but solvable equation of prime degree are rationally expressible in terms of any two of them.**

For the identical permutation is the only permutation of the metacyclic group leaving two roots x_i and x_k unaltered, which is readily verified from the nature of such permutations:

$$t = \binom{z}{\mu z + \nu}.$$

Hence the function

$$v = \alpha_i x_i + \alpha_k x_k$$

belongs to the group identity, and every root of the equation is rationally expressible in terms of v, which is to say in terms of x_i and x_k.

Conversely:

(152) **An irreducible equation of prime degree whose roots are rationally expressible in terms of any two of them is solvable.**

For the group G of the equation is transitive on the roots

$$x_0, x_1, \ . \ . \ . \ , x_{p-1}$$

of the equation,[1] whence its order is a multiple of p.[2] But p is prime; by the Theorem of Cauchy the group G has a subgroup of order p, and by proposition (25) this subgroup is circular.

Let the subgroup be C formed by the powers of

$$s = (01 \ . \ . \ . \ p - 1),$$

where 0 and 1 designate any two roots of the proposition. Unless the subgroup C is normal in the group G, there is in G a permutation

$$t = (i_0 i_1 \ . \ . \ . \ i_{p-1})$$

similar to s yet not a power of s, by proposition (30).

If μ and ν are two numbers not congruent modulo p, we apply to the relation

$$x_{i_\alpha} = R(x_{i_0}, x_{i_1})$$

[1] Cf. proposition (123).

[2] Cf. proposition (60).

of the proposition the permutations $t^\mu s^{-i_\mu}$ and $t^\nu s^{-i_\nu}$ of G and obtain the two relations

$$x_{i_{\alpha+\mu}-i_\mu} = R(x_0, x_{i_{1+\mu}-i_\mu})$$
$$x_{i_{\alpha+\nu}-i_\nu} = R(x_0, x_{i_{1+\nu}-i_\nu}),$$

which are true by proposition (107).

The numbers μ and ν can be chosen so that

$$i_{\mu+1} - i_\mu \equiv i_{\nu+1} - i_\nu \equiv k(\text{mod } p),$$

for there are p such differences and they can take only $p - 1$ values incongruent modulo p, zero not being possible. This gives

$$x_{i_{\alpha+\mu}-i_\mu} = R(x_0, x_k)$$
$$x_{i_{\alpha+\nu}-i_\nu} = R(x_0, x_k);$$

whence

$$i_{\alpha+\mu} - i_\mu = i_{\alpha+\nu} - i_\nu$$

for

$$\alpha = 0, 1, \ldots, p - 1.$$

If these relations are written out in full, it is readily seen that they imply the other:

$$i_{\alpha+1} - i_\alpha = \text{constant},$$

say m. But this means that

$$t = s^m$$

contrary to our assumption; hence C is normal in G. Since the largest group containing C as normal subgroup is metacyclic, the proposition follows.

The roots of an equation being rationally expressible in terms of any two of them thus is a necessary and sufficient condition by which we recognize the equation as solvable when it is irreducible and of prime degree.

It appears that

(153) **an irreducible but solvable equation with real coefficients whose degree is an odd prime has only one real root or real roots alone.**

For it has one real root by a proposition of elementary algebra; and if two roots are real, all are so by proposition (151).

Irreducible but solvable equations of prime degree have an other peculiar property:

(154) **An irreducible equation of prime degree is solvable when, and only when, the general resolvent[1] for the metacyclic group having no double roots has one rational root.**

Such a resolvent for the general equation is

$$r(\psi) = (\psi - \psi_1)(\psi - \psi_2) \ . \ . \ . = 0,$$

where ψ_1 is a function belonging to the metacyclic group. Its degree is equal to the index of the metacyclic group in the symmetric, which is

$$j_m = 1 . 2 \ . \ . \ . \ (p - 2);$$

and it is irreducible with coefficients that are symmetric by proposition (119).

Suppose now we substitute for the roots x_i of the general equation those of the special equation given by the proposition. Let no two ψ_i become alike, which implies that ψ_1 still belongs to the metacyclic group; this can always be arranged by a proper choice of ψ_1.

If the group of the equation is soluble, the root ψ_1 of the resolvent has a rational value by property (2) of the group, because ψ_1 is unaltered by the largest possible such group, the metacyclic.

Conversely, if the root ψ_1 of the resolvent has a rational value, the group of the equation is contained in the metacyclic by property (1), because ψ_1 is unaltered by the metacyclic group. Hence the proposition follows.

To the irreducible but solvable equations of prime degree belong the binomial equations. For

(155) **a binomial equation of prime degree in Ω with no root rational in Ω is irreducible there but solvable.**

Let such an equation be

$$x^p - c = 0;$$

it has the roots

$$x_0, \ \epsilon x_0, \ . \ . \ . \ , \ \epsilon^{p-1} x_0$$

when x_0 is one of them and ϵ a primitive root of unity.

Suppose that

$$x^p - c$$

[1] We mean the resolvent for the general equation with the roots of the given equation substituted.

has a factor

$$x^q + \dots + d$$

in Ω with

$$q < p.$$

The product of its roots, which is to say the constant term

$$d = \epsilon^z x_0{}^q$$

with unknown z, is rational in Ω.

By proposition (3) which applies also to integers,[1] there exist two integers a and b such that

$$aq - bp = 1.$$

Hence rational in Ω is also

$$(\epsilon^z x_0{}^q)^a = \epsilon^{az} x_0{}^{bp+1} = \epsilon^{az} x_0 \,.\, c^b,$$

where $\epsilon^{az} x_0$ is a root of the equation.

With c^b also $\epsilon^{az} x_0$ then is rational in Ω, which is contrary to the assumption of our proposition.

Writing the roots of the equation in the form

$$x_0,\ x_1,\ \dots,\ x_{p-1},$$

we have

$$\epsilon = \frac{x_1}{x_0}$$

and

$$x_i = \left(\frac{x_1}{x_0}\right)^i x_0.$$

It appears that all roots of the equation are rationally expressible in terms of two of them, which completes the proof.

As function belonging to the cyclic group

$$C = \{s\}$$

with

$$s = (01 \dots p-1)$$

we can take

$$\epsilon = \frac{x_1}{x_0} = \frac{x_2}{x_1} = \dots = \frac{x_p}{x_{p-1}}.$$

But ϵ satisfies the cyclotomic equation

$$x^{p-1} + \dots + x + 1 = 0$$

of degree $p - 1$. In the domain of rational numbers this equation is irreducible by proposition (146) and serves as resolvent

[1] Compare proposition (91).

of the binomial equation, whence the group of the binomial equation is metacyclic containing C as subgroup of index $p - 1$. We conclude that

(156) **a binomial equation of prime degree with no rational root is metacyclic in the domain of rational numbers.**

§86. QUINTIC EQUATION

The general quintic has a resolvent of degree six, as already Lagrange knew; for the metacyclic group is of index six in the symmetric. And when the resolvent for the metacyclic group takes for a special quintic one rational root, the quintic is solvable by proposition (154). It is possible to construct such a resolvent sextic, but to compute its coefficients is very painful. Somewhat lighter is the task when the quintic is reduced to the **Bring-Jerrard normal form**

$$x^5 + px + q = 0.$$

To reduce the quintic, we first prove that

(157) **every quadratic form is expressible as sum of squares of linear forms.**

By a **form** we mean a homogeneous function, by a quadratic form

$$\varphi = \varphi(x_i)_1^n$$

a homogeneous function of degree two in the x_i. If it contains terms in $x_1{}^2$, we can set

$$\varphi = c^2x_1{}^2 + 2c\psi x_1 + \omega,$$

where ψ is linear and ω quadratic in the x_i other than x_1. But this is the same as

$$\begin{aligned}\varphi &= (cx_1 + \psi)^2 + (\omega - \psi^2)\\ &= u^2 + (\omega - \psi^2),\end{aligned}$$

where $\omega - \psi^2$ is a quadratic form in the x_i other than x_1.

If φ contains no square of an x_i, we can set

$$\varphi = c^2x_1x_2 + c\chi x_1 + c\psi x_2 + \omega,$$

where χ and ψ are linear and ω is quadratic in the x_i other than x_1 and x_2. But this is the same as

$$\begin{aligned}\varphi = \tfrac{1}{4}[c(x_1 + x_2) + (\chi + \psi)]^2 - \tfrac{1}{4}[c(x_1 - x_2) - (\chi - \psi)]^2\\ + (\omega - \chi\psi) = \tfrac{1}{4}u^2 - \tfrac{1}{4}v^2 + (\omega - \chi\psi),\end{aligned}$$

where $\omega - \chi\psi$ is a quadratic form in the x_i other than x_1 and x_2.

The quadratic forms obtained in fewer than n letters x_i we then treat as we treated φ and continue until φ is expressed as sum of squared linear forms. Evidently there can be no more than n such squares.

Jerrard proved, what Bring achieved much earlier for a quintic equation, that

(158) **by a Tschirnhaus transformation involving square and cube roots the second, third and fourth terms of an equation can be removed.**

The general equation

$$x^n + a_1x^{n-1} + \ldots + a_n = 0$$

is transformed into

$$y^n + b_1y^{n-1} + \ldots + b_n = 0$$

by the substitution

$$y = \alpha_0 + \alpha_1x + \alpha_2x^2 + \alpha_3x^3 + \alpha_4x^4,$$

which is a case of the **Tschirnhaus transformation**

$$y = \frac{\varphi(x)}{\psi(x)}.$$

For this substitution we have

$$y^2 = \beta_0 + \beta_1x + \ldots + \beta_{n-1}x^{n-1}$$

$$\cdots\cdots\cdots$$

$$y^n = \delta_0 + \delta_1x + \ldots + \delta_{n-1}x^{n-1},$$

since powers of x higher than $n - 1$ can be reduced from[1]

$$x^n = -a_1x^{n-1} - \ldots - a_n.$$

Any β_i is homogeneous of degree two, any δ_i is homogeneous of degree n in the α_i.

The relations between x and y hold for any x_i of the general equation, for every x_i we have a y_i, and setting

$$s_i = s(x^i)$$
$$\sigma_i = s(y^i)$$

in the notation of §22 with the subscripts of x and y not marked, we find

$$\sigma_1 = n\alpha_0 + \alpha_1s_1 + \alpha_2s_2 + \alpha_3s_3 + \alpha_4s_4$$
$$\sigma_2 = n\beta_0 + \beta_1s_1 + \ldots + \beta_{n-1}s_{n-1}$$
$$\cdots\cdots\cdots$$

[1] Compare the same for ψ in §35.

The s_i are computable from the a_i, by proposition (34); likewise the σ_i are computable from the b_i:

$$\sigma_1 = -b_1$$
$$\sigma_2 = b_1^2 - 2b_2$$
$$\cdot \quad \cdot \quad \cdot \quad \cdot \quad \cdot \quad \cdot \quad \cdot$$

Any σ_i being homogeneous of degree i in the α_i, the same is true for any b_i.

From the equation

$$b_1 = 0$$

linear in the α_i we obtain one α_i in terms of the other four and eliminate it from the equations

$$b_2 = 0$$
$$b_3 = 0,$$

putting the equation for b_2 by proposition (157) in the form

$$\lambda_1 u_1^2 + \lambda_2 u_2^2 + \lambda_3 u_3^2 + \lambda_4 u_4^2 = 0,$$

where any u_i is linear in the α_i.

Then we find such α_i that

$$\lambda_1 u_1^2 + \lambda_2 u_2^2 = 0$$
$$\lambda_3 u_3^2 + \lambda_4 u_4^2 = 0,$$

or

$$\sqrt{\lambda_1}\ u_1 = \sqrt{-\lambda_2}\ u_2$$
$$\sqrt{\lambda_3}\ u_3 = \sqrt{-\lambda_4}\ u_4,$$

which is to say express two α_i in terms of the remaining two.

Substituting into the equation

$$b_3 = 0,$$

we have an equation homogeneous of degree three in two α_i. Choosing one α_i at will, we solve for the other, and all α_i are determined.

With the α_i the substitution is determined which removes from the general equation the terms in y^{n-1}, y^{n-2}, y^{n-3}.

Thus we can take the general quintic

$$x^5 + a_1x^4 + a_2x^3 + a_3x^2 + a_4x + a_5 = 0$$

to be

$$x^5 + px + q = 0,$$

with

$$a_1 = a_2 = a_3 = 0$$
$$a_4 = p,\ a_5 = q.$$

The function ψ of its roots given in §58 and belonging to the half-metacyclic group takes under the alternating the conjugate values:[1]

$$\begin{aligned}
\psi_1 &= x_1x_2 + x_2x_3 + x_3x_4 + x_4x_5 + x_5x_1 \\
\psi_2 &= x_2x_1 + x_1x_4 + x_4x_3 + x_3x_5 + x_5x_2 \\
\psi_3 &= x_2x_4 + x_4x_5 + x_5x_3 + x_3x_1 + x_1x_2 \\
\psi_4 &= x_5x_4 + x_4x_1 + x_1x_3 + x_3x_2 + x_2x_5 \\
\psi_5 &= x_2x_4 + x_4x_1 + x_1x_5 + x_5x_3 + x_3x_2 \\
\psi_6 &= x_2x_5 + x_5x_1 + x_1x_3 + x_3x_4 + x_4x_2.
\end{aligned}$$

The function ψ' conjugate to ψ in the metacyclic group takes under the alternating the corresponding values:

$$\begin{aligned}
\psi_1' &= x_1x_3 + x_1x_4 + x_2x_4 + x_2x_5 + x_3x_5 \\
\psi_2' &= x_1x_3 + x_1x_5 + x_2x_3 + x_2x_4 + x_4x_5 \\
\psi_3' &= x_1x_4 + x_1x_5 + x_2x_3 + x_2x_5 + x_3x_4 \\
\psi_4' &= x_1x_2 + x_1x_5 + x_2x_4 + x_3x_4 + x_3x_5 \\
\psi_5' &= x_1x_2 + x_1x_3 + x_2x_5 + x_3x_4 + x_4x_5 \\
\psi_6' &= x_1x_2 + x_1x_4 + x_2x_3 + x_3x_5 + x_4x_5
\end{aligned}$$

such that

$$\psi_i + \psi_i' = a_2.$$

The function

$$\omega_1 = \psi_1 - \psi_1'$$

belongs to the half-metacyclic group; the six conjugate values of ω_1 under the alternating group are roots of the resolvent

$$(\omega - \omega_1)(\omega - \omega_2)(\omega - \omega_3)(\omega - \omega_4)(\omega - \omega_5)(\omega - \omega_6) = 0$$

whose coefficients are unaltered by the group, by proposition (50). Since the ω_i change their sign when the root of the discriminant does, the resolvent is:

$$\omega^6 + \lambda_2\omega^4 + \lambda_4\omega^2 + \lambda_6 - \sqrt{\Delta}(\lambda_1\omega^5 + \lambda_3\omega^3 + \lambda_5\omega) = 0,$$

where the λ_i are symmetric in the x_i.

We find for

$$\lambda_1\sqrt{\Delta},\ \lambda_2,\ \lambda_3\sqrt{\Delta},\ \lambda_4,\ \lambda_5\sqrt{\Delta},\ \lambda_6$$

the total degrees

$$2 \quad 4 \quad 6 \quad 8 \quad 10 \quad 12$$

in the x_i. But $\sqrt{\Delta}$ is of total degree ten, whence λ_5 is some number and

$$\lambda_1 = \lambda_3 = 0.$$

[1] We apply to ψ_1 the permutations: (12)(34), (1243)(15) = (12435) (1243)(25) = (15243), (1243)(35) = (12453), (1243)(45) = (12543).

The weight for

$$\lambda_2, \lambda_4, \lambda_6$$

in the a_i is

$$4 \quad 8 \quad 12,$$

by §24. Therefore a_5 is not contained in λ_2; it is not contained in λ_4, λ_6 either because it could be there only in combination with a_1, a_2, a_3 which are zero.

Consequently λ_2, λ_4, λ_6 are not affected if we assume that

$$q = 0.$$

Then

$$x_1 = 0,\ x_2 = (-p)^{1/4},\ x_3 = i(-p)^{1/4}$$
$$x_4 = -(-p)^{1/4},\ x_5 = -i(-p)^{1/4},$$

whence

$$\sqrt{\Delta} = \prod_{i<k}(x_i - x_k) = -16\sqrt{p^5}$$

$$\Delta = 256\, p^5.$$

From

$$\psi_i + \psi_i' = 0$$
$$\psi_i - \psi_i' = \omega_i$$

we have

$$\omega_i = 2\psi_i,$$

and substituting the values for the x_i:

$$\omega_1 = \omega_3 = \omega_4 = \omega_5 = -2\sqrt{p}$$
$$\omega_2 = (4 + 2i)\sqrt{p},\ \omega_6 = (4 - 2i)\sqrt{p}.$$

The resolvent for this case is

$$(\omega + 2\sqrt{p})^4(\omega^2 - 8\omega\sqrt{p} + 20p)$$
$$= \omega^6 - 20\, p\omega^4 + 240\, p^2\omega^2 + 512\sqrt{p^5}\omega + 320\, p^3 = 0.$$

Hence the resolvent is always

$$\omega^6 - 20\, p\omega^4 + 240\, p^2\omega^2 - 32\sqrt{\Delta}\,\omega + 320\, p^3 = 0,$$

considering that λ_5 does not depend on q.

The discriminant is expressible in terms of p and q. Being of total degree twenty in the x_i, it is

$$\Delta = kp^5 + lq^4.$$

We find k by setting $q = 0$ and l by setting $p = 0$ and then have

$$\Delta = 2^8p^5 + 5^5q^4.$$

In terms of ψ the resolvent is

$$\psi^6 - 5p\psi^4 + 15p^2\psi^2 - \sqrt{\Delta}\,\psi + 5p^3 = 0.$$

Substituting

$$\varphi_i = \psi_i^2$$

we find:

(159) The Bring-Jerrard normal form

$$\boxed{x^5 + px + q = 0}$$

of the general quintic has a resolvent sextic

$$\boxed{(\varphi^3 - 5p\varphi^2 + 15p^2\varphi + 5p^3)^2 = \Delta\varphi}$$

for the metacyclic group. It is the resolvent also of a special quintic if the special quintic is irreducible.

For the degree of the resolvent is six, its coefficients are symmetric and its root φ_1 belongs to the metacyclic group.

When the roots x_i of a special quintic are substituted, φ_1 still belongs to the metacyclic group if the quintic is irreducible, because then no φ_i are alike.

The resolvent in φ has equal roots only when the resolvent in ψ has equal roots or roots differing by the sign alone.

Two roots ψ_i differing by the sign alone are possible only when the resolvent in ψ has

$$\Delta = 0,$$

which is to say when the quintic has double roots. But this case is excluded.

Equal roots ψ_i are possible only when the resolvent in ψ has a root in common with

$$6\psi^5 - 20p\psi^3 + 30p^2\psi - \sqrt{\Delta} = 0,$$

or eliminating $\sqrt{\Delta}$:

$$5\psi^6 - 15p\psi^4 + 15p^2\psi^2 - 5p^3 = 5(\psi^2 - p)^3 = 0,$$

which is to say when

$$\varphi = p.$$

But this means that

$$q = 0,$$

which is readily verified from the resolvent in φ with the value of Δ put in. In this case the quintic is reducible, which proves the proposition.

Hence it follows that any quintic in the Bring-Jerrard normal form is solvable when, and only when, it is irreducible and the general resolvent for the metacyclic group has a rational root.

INDEX

Catalogue of Dover
SCIENCE BOOKS

BOOKS THAT EXPLAIN SCIENCE

THE NATURE OF LIGHT AND COLOUR IN THE OPEN AIR, M. Minnaert. Why is falling snow sometimes black? What causes mirages, the fata morgana, multiple suns and moons in the sky; how are shadows formed? Prof. Minnaert of U. of Utrecht answers these and similar questions in optics, light, colour, for non-specialists. Particularly valuable to nature, science students, painters, photographers. "Can best be described in one word—fascinating!" Physics Today. Translated by H. M. Kremer-Priest, K. Jay. 202 illustrations, including 42 photos. xvi + 362pp. 5⅜ x 8. T196 Paperbound **$1.95**

THE RESTLESS UNIVERSE, Max Born. New enlarged version of this remarkably readable account by a Nobel laureate. Moving from sub-atomic particles to universe, the author explains in very simple terms the latest theories of wave mechanics. Partial contents: air and its relatives, electrons and ions, waves and particles, electronic structure of the atom, nuclear physics. Nearly 1000 illustrations, including 7 animated sequences. 325pp. 6 x 9. T412 Paperbound **$2.00**

MATTER AND LIGHT, THE NEW PHYSICS, L. de Broglie. Non-technical papers by a Nobel laureate explain electromagnetic theory, relativity, matter, light, radiation, wave mechanics, quantum physics, philosophy of science. Einstein, Planck, Bohr, others explained so easily that no mathematical training is needed for all but 2 of the 21 chapters. "Easy simplicity and lucidity . . . should make this source-book of modern physcis available to a wide public," Saturday Review. Unabridged. 300pp. 5⅜ x 8. T35 Paperbound **$1.60**

THE COMMON SENSE OF THE EXACT SCIENCES, W. K. Clifford. Introduction by James Newman, edited by Karl Pearson. For 70 years this has been a guide to classical scientific, mathematical thought. Explains with unusual clarity basic concepts such as extension of meaning of symbols, characteristics of surface boundaries, properties of plane figures, vectors, Cartesian method of determining position, etc. Long preface by Bertrand Russell. Bibliography of Clifford. Corrected. 130 diagrams redrawn. 249pp. 5⅜ x 8. T61 Paperbound **$1.60**

THE EVOLUTION OF SCIENTIFIC THOUGHT FROM NEWTON TO EINSTEIN, A. d'Abro. Einstein's special, general theories of relativity, with historical implications, analyzed in non-technical terms. Excellent accounts of contributions of Newton, Riemann, Weyl, Planck, Eddington, Maxwell, Lorentz, etc., are treated in terms of space, time, equations of electromagnetics, finiteness of universe, methodology of science. "Has become a standard work," Nature. 21 diagrams. 482pp. 5⅜ x 8. T2 Paperbound **$2.00**

BRIDGES AND THEIR BUILDERS, D. Steinman, S. R. Watson. Engineers, historians, everyone ever fascinated by great spans will find this an endless source of information and interest. Dr. Steinman, recent recipient of Louis Levy Medal, is one of the great bridge architects, engineers of all time. His analysis of great bridges of history is both authoritative and easily followed. Greek, Roman, medieval, oriental bridges; modern works such as Brooklyn Bridge, Golden Gate Bridge, etc. described in terms of history, constructional principles, artistry, function. Most comprehensive, accurate semi-popular history of bridges in print in English. New, greatly revised, enlarged edition. 23 photographs, 26 line drawings. xvii + 401pp. 5⅜ x 8. T431 Paperbound **$1.95**

CONCERNING THE NATURE OF THINGS, Sir William Bragg. Christmas lectures at Royal Society by Nobel laureate, dealing with atoms, gases, liquids, and various types of crystals. No scientific background is needed to understand this remarkably clear introduction to basic processes and aspects of modern science. "More interesting than any bestseller," London Morning Post. 32pp. of photos. 57 figures. xii + 232pp. 5⅜ x 8. T31 Paperbound **$1.35**

THE RISE OF THE NEW PHYSICS, A. d'Abro. Half million word exposition, formerly titled "The Decline of Mechanism," for readers not versed in higher mathematics. Only thorough explanation in everyday language of core of modern mathematical physical theory, treating both classical, modern views. Scientifically impeccable coverage of thought from Newtonian system through theories of Dirac, Heisenberg, Fermi's statistics. Combines history, exposition; broad but unified, detailed view, with constant comparison of classical, modern views. "A must for anyone doing serious study in the physical sciences," J. of the Franklin Inst. "Extraordinary faculty . . . to explain ideas and theories . . . in language of everyday life," Isis. Part I of set: philosophy of science, from practice of Newton, Maxwell, Poincaré, Einstein, etc. Modes of thought, experiment, causality, etc. Part II: 100 pp. on grammar, vocabulary of mathematics, discussions of functions, groups, series, Fourier series, etc. Remainder treats concrete, detailed coverage of both classical, quantum physics: analytic mechanics, Hamilton's principle, electromagnetic waves, thermodynamics, Brownian movement, special relativity, Bohr's atom, de Broglie's wave mechanics, Heisenberg's uncertainty, scores of other important topics. Covers discoveries, theories of d'Alembert, Born, Cantor, Debye, Euler, Foucault, Galois, Gauss, Hadamard, Kelvin, Kepler Laplace, Maxwell, Pauli, Rayleigh Volterra, Weyl, more than 180 others. 97 illustrations. ix + 982pp. 5⅜ x 8.
T3 Vol. 1 Paperbound **$2.00**
T4 Vol. II Paperbound **$2.00**

SPINNING TOPS AND GYROSCOPIC MOTION, John Perry. Well-known classic of science still unsurpassed for lucid, accurate, delightful exposition. How quasi-rigidity is induced in flexible, fluid bodies by rapid motions; why gyrostat falls, top rises; nature, effect of internal fluidity on rotating bodies; etc. Appendixes describe practical use of gyroscopes in ships, compasses, monorail transportation. 62 figures. 128pp. 5⅜ x 8.
T416 Paperbound **$1.00**

FOUNDATIONS OF PHYSICS, R. B. Lindsay, H. Margenau. Excellent bridge between semi-popular and technical writings. Discussion of methods of physical description, construction of theory; valuable to physicist with elementary calculus. Gives meaning to data, tools of modern physics. Contents: symbolism, mathematical equations; space and time; foundations of mechanics; probability; physics, continua; electron theory; relativity; quantum mechanics; causality; etc. "Thorough and yet not overdetailed. Unreservedly recommended," Nature. Unabridged corrected edition. 35 illustrations. xi + 537pp. 5⅜ x 8. S377 Paperbound **$2.45**

FADS AND FALLACIES IN THE NAME OF SCIENCE, Martin Gardner. Formerly entitled "In the Name of Science," the standard account of various cults, quack systems, delusions which have masqueraded as science: hollow earth fanatics, orgone sex energy, dianetics, Atlantis, Forteanism, flying saucers, medical fallacies like zone therapy, etc. New chapter on Bridey Murphy, psionics, other recent manifestations. A fair reasoned appraisal of eccentric theory which provides excellent innoculation. "Should be read by everyone, scientist or non-scientist alike," R. T. Birge, Prof. Emeritus of Physics, Univ. of Calif; Former Pres., Amer. Physical Soc. x + 365pp. 5⅜ x 8. T394 Paperbound **$1.50**

ON MATHEMATICS AND MATHEMATICIANS, R. E. Moritz. A 10 year labor of love by discerning, discriminating Prof. Moritz, this collection conveys the full sense of mathematics and personalities of great mathematicians. Anecdotes, aphorisms, reminiscences, philosophies, definitions, speculations, biographical insights, etc. by great mathematicians, writers: Descartes, Mill, Locke, Kant, Coleridge, Whitehead, etc. Glimpses into lives of great mathematicians, from Archimedes to Euler, Gauss, Weierstrass. To mathematicians, a superb browsing-book. To laymen, exciting revelation of fullness of mathematics. Extensive cross index. 410pp. 5⅜ x 8. T489 Paperbound **$1.95**

GUIDE TO THE LITERATURE OF MATHEMATICS AND PHYSICS, N. G. Parke III. Over 5000 entries under approximately 120 major subject headings, of selected most important books, monographs, periodicals, articles in English, plus important works in German, French, Italian, Spanish, Russian (many recently available works). Covers every branch of physics, math, related engineering. Includes author, title, edition, publisher, place, date, number of volumes, number of pages. 40 page introduction on basic problems of research, study provides useful information on organization, use of libraries, psychology of learning, etc. Will save you hours of time. 2nd revised edition. Indices of authors, subjects. 464pp. 5⅜ x 8. S447 Paperbound **$2.49**

THE STRANGE STORY OF THE QUANTUM, An Account for the General Reader of the Growth of Ideas Underlying Our Present Atomic Knowledge, B. Hoffmann. Presents lucidly, expertly, with barest amount of mathematics, problems and theories which led to modern quantum physics. Begins with late 1800's when discrepancies were noticed; with illuminating analogies, examples, goes through concepts of Planck, Einstein, Pauli, Schroedinger, Dirac, Sommerfield, Feynman, etc. New postscript through 1958. "Of the books attempting an account of the history and contents of modern atomic physics which have come to my attention, this is the best," H. Margenau, Yale U., in Amer. J. of Physics. 2nd edition. 32 tables, illustrations. 275pp. 5⅜ x 8. T518 Paperbound **$1.45**

HISTORY OF SCIENCE AND PHILOSOPHY OF SCIENCE

THE VALUE OF SCIENCE, Henri Poincaré. Many of most mature ideas of "last scientific universalist" for both beginning, advanced workers. Nature of scientific truth, whether order is innate in universe or imposed by man, logical thought vs. intuition (relating to Weierstrass, Lie, Riemann, etc), time and space (relativity, psychological time, simultaneity), Herz's concept of force, values within disciplines of Maxwell, Carnot, Mayer, Newton, Lorentz, etc. iii + 147pp. 5⅜ x 8. S469 Paperbound **$1.35**

PHILOSOPHY AND THE PHYSICISTS, L. S. Stebbing. Philosophical aspects of modern science examined in terms of lively critical attack on ideas of Jeans, Eddington. Tasks of science, causality, determinism, probability, relation of world physics to that of everyday experience, philosophical significance of Planck-Bohr concept of discontinuous energy levels, inferences to be drawn from Uncertainty Principle, implications of "becoming" involved in 2nd law of thermodynamics, other problems posed by discarding of Laplacean determinism. 285pp. 5⅜ x 8. T480 Paperbound **$1.65**

THE PRINCIPLES OF SCIENCE, A TREATISE ON LOGIC AND THE SCIENTIFIC METHOD, W. S. Jevons. Milestone in development of symbolic logic remains stimulating contribution to investigation of inferential validity in sciences. Treats inductive, deductive logic, theory of number, probability, limits of scientific method; significantly advances Boole's logic, contains detailed introduction to nature and methods of probability in physics, astronomy, everyday affairs, etc. In introduction, Ernest Nagel of Columbia U. says, "[Jevons] continues to be of interest as an attempt to articulate the logic of scientific inquiry." liii + 786pp. 5⅜ x 8. S446 Paperbound **$2.98**

A HISTORY OF ASTRONOMY FROM THALES TO KEPLER, J. L. E. Dreyer. Only work in English to give complete history of cosmological views from prehistoric times to Kepler. Partial contents: Near Eastern astronomical systems, Early Greeks, Homocentric spheres of Euxodus, Epicycles, Ptolemaic system, Medieval cosmology, Copernicus, Kepler, much more. "Especially useful to teachers and students of the history of science . . . unsurpassed in its field," Isis. Formerly "A History of Planetary Systems from Thales to Kepler." Revised foreword by W. H. Stahl. xvii + 430pp. 5⅜ x 8. S79 Paperbound **$1.98**

A CONCISE HISTORY OF MATHEMATICS, D. Struik. Lucid study of development of ideas, techniques, from Ancient Near East, Greece, Islamic science, Middle Ages, Renaissance, modern times. Important mathematicians described in detail. Treatment not anecdotal, but analytical development of ideas. Non-technical—no math training needed. "Rich in content, thoughtful in interpretations," U.S. Quarterly Booklist. 60 illustrations including Greek, Egyptian manuscripts, portraits of 31 mathematicians. 2nd edition. xix + 299pp. 5⅜ x 8. S255 Paperbound **$1.75**

THE PHILOSOPHICAL WRITINGS OF PEIRCE, edited by Justus Buchler. A carefully balanced expositon of Peirce's complete system, written by Peirce himself. It covers such matters as scientific method, pure chance vs. law, symbolic logic, theory of signs, pragmatism, experiment, and other topics. "Excellent selection . . . gives more than adequate evidence of the range and greatness," Personalist. Formerly entitled "The Philosophy of Peirce." xvi + 368pp. T217 Paperbound **$1.95**

SCIENCE AND METHOD, Henri Poincaré. Procedure of scientific discovery, methodology, experiment, idea-germination—processes by which discoveries come into being. Most significant and interesting aspects of development, application of ideas. Chapters cover selection of facts, chance, mathematical reasoning, mathematics and logic; Whitehead, Russell, Cantor, the new mechanics, etc. 288pp. 5⅜ x 8. S222 Paperbound **$1.35**

SCIENCE AND HYPOTHESIS, Henri Poincaré. Creative psychology in science. How such concepts as number, magnitude, space, force, classical mechanics developed, how modern scientist uses them in his thought. Hypothesis in physics, theories of modern physics. Introduction by Sir James Larmor. "Few mathematicians have had the breadth of vision of Poincaré, and none is his superior in the gift of clear exposition," E. T. Bell. 272pp. 5⅜ x 8. S221 Paperbound **$1.35**

ESSAYS IN EXPERIMENTAL LOGIC, John Dewey. Stimulating series of essays by one of most influential minds in American philosophy presents some of his most mature thoughts on wide range of subjects. Partial contents: Relationship between inquiry and experience; dependence of knowledge upon thought; character logic; judgments of practice, data, and meanings; stimuli of thought, etc. viii + 444pp. 5⅜ x 8. T73 Paperbound **$1.95**

WHAT IS SCIENCE, Norman Campbell. Excellent introduction explains scientific method, role of mathematics, types of scientific laws. Contents: 2 aspects of science, science and nature, laws of chance, discovery of laws, explanation of laws, measurement and numerical laws, applications of science. 192pp. 5⅜ x 8. S43 Paperbound **$1.25**

FROM EUCLID TO EDDINGTON: A STUDY OF THE CONCEPTIONS OF THE EXTERNAL WORLD, Sir Edmund Whittaker. Foremost British scientist traces development of theories of natural philosophy from western rediscovery of Euclid to Eddington, Einstein, Dirac, etc. 5 major divisions: Space, Time and Movement; Concepts of Classical Physics; Concepts of Quantum Mechanics; Eddington Universe. Contrasts inadequacy of classical physics to understand physical world with present day attempts of relativity, non-Euclidean geometry, space curvature, etc. 212pp. 5⅜ x 8. T491 Paperbound **$1.35**

THE ANALYSIS OF MATTER, Bertrand Russell. How do our senses accord with the new physics? This volume covers such topics as logical analysis of physics, prerelativity physics, causality, scientific inference, physics and perception, special and general relativity, Weyl's theory, tensors, invariants and their physical interpretation, periodicity and qualitative series. "The most thorough treatment of the subject that has yet been published," The Nation. Introduction by L. E. Denonn. 422pp. 5⅜ x 8. T231 Paperbound **$1.95**

LANGUAGE, TRUTH, AND LOGIC, A. Ayer. A clear introduction to the Vienna and Cambridge schools of Logical Positivism. Specific tests to evaluate validity of ideas, etc. Contents: function of philosophy, elimination of metaphysics, nature of analysis, a priori, truth and probability, etc. 10th printing. "I should like to have written it myself," Bertrand Russell. 160pp. 5⅜ x 8. T10 Paperbound **$1.25**

THE PSYCHOLOGY OF INVENTION IN THE MATHEMATICAL FIELD, J. Hadamard. Where do ideas come from? What role does the unconscious play? Are ideas best developed by mathematical reasoning, word reasoning, visualization? What are the methods used by Einstein, Poincaré, Galton, Riemann? How can these techniques be applied by others? One of the world's leading mathematicians discusses these and other questions. xiii + 145pp. 5⅜ x 8. T107 Paperbound **$1.25**

GUIDE TO PHILOSOPHY, C. E. M. Joad. By one of the ablest expositors of all time, this is not simply a history or a typological survey, but an examination of central problems in terms of answers afforded by the greatest thinkers: Plato, Aristotle, Scholastics, Leibniz, Kant, Whitehead, Russell, and many others. Especially valuable to persons in the physical sciences; over 100 pages devoted to Jeans, Eddington, and others, the philosophy of modern physics, scientific materialism, pragmatism, etc. Classified bibliography. 592pp. 5⅜ x 8. T50 Paperbound **$2.00**

SUBSTANCE AND FUNCTION, and **EINSTEIN'S THEORY OF RELATIVITY, Ernst Cassirer.** Two books bound as one. Cassirer establishes a philosophy of the exact sciences that takes into consideration new developments in mathematics, shows historical connections. Partial contents: Aristotelian logic, Mill's analysis, Helmholtz and Kronecker, Russell and cardinal numbers, Euclidean vs. non-Euclidean geometry, Einstein's relativity. Bibliography. Index. xxi + 464pp. 5⅜ x 8. T50 Paperbound **$2.00**

FOUNDATIONS OF GEOMETRY, Bertrand Russell. Nobel laureate analyzes basic problems in the overlap area between mathematics and philosophy: the nature of geometrical knowledge, the nature of geometry, and the applications of geometry to space. Covers history of non-Euclidean geometry, philosophic interpretations of geometry, especially Kant, projective and metrical geometry. Most interesting as the solution offered in 1897 by a great mind to a problem still current. New introduction by Prof. Morris Kline, N.Y. University. "Admirably clear, precise, and elegantly reasoned analysis," International Math. News. xii + 201pp. 5⅜ x 8. S233 Paperbound **$1.60**

THE NATURE OF PHYSICAL THEORY, P. W. Bridgman. How modern physics looks to a highly unorthodox physicist—a Nobel laureate. Pointing out many absurdities of science, demonstrating inadequacies of various physical theories, weighs and analyzes contributions of Einstein, Bohr, Heisenberg, many others. A non-technical consideration of correlation of science and reality. xi + 138pp. 5⅜ x 8. S33 Paperbound **$1.25**

EXPERIMENT AND THEORY IN PHYSICS, Max Born. A Nobel laureate examines the nature and value of the counterclaims of experiment and theory in physics. Synthetic versus analytical scientific advances are analyzed in works of Einstein, Bohr, Heisenberg, Planck, Eddington, Milne, others, by a fellow scientist. 44pp. 5⅜ x 8. S308 Paperbound **60¢**

A SHORT HISTORY OF ANATOMY AND PHYSIOLOGY FROM THE GREEKS TO HARVEY, Charles Singer. Corrected edition of "The Evolution of Anatomy." Classic traces anatomy, physiology from prescientific times through Greek, Roman periods, dark ages, Renaissance, to beginning of modern concepts. Centers on individuals, movements, that definitely advanced anatomical knowledge. Plato, Diocles, Erasistratus, Galen, da Vinci, etc. Special section on Vesalius. 20 plates. 270 extremely interesting illustrations of ancient, Medieval, enaissance, Oriental origin. xii + 209pp. 5⅜ x 8. T389 Paperbound **$1.75**

SPACE-TIME-MATTER, Hermann Weyl. "The standard treatise on the general theory of relativity," (Nature), by world renowned scientist. Deep, clear discussion of logical coherence of general theory, introducing all needed tools: Maxwell, analytical geometry, non-Euclidean geometry, tensor calculus, etc. Basis is classical space-time, before absorption of relativity. Contents: Euclidean space, mathematical form, metrical continuum, general theory, etc. 15 diagrams. xviii + 330pp. 5⅜ x 8. S267 Paperbound **$1.75**

DOVER SCIENCE BOOKS

MATTER AND MOTION, James Clerk Maxwell. Excellent exposition begins with simple particles, proceeds gradually to physical systems beyond complete analysis; motion, force, properties of centre of mass of material system; work, energy, gravitation, etc. Written with all Maxwell's original insights and clarity. Notes by E. Larmor. 17 diagrams. 178pp. 5⅜ x 8. S188 Paperbound **$1.25**

PRINCIPLES OF MECHANICS, Heinrich Hertz. Last work by the great 19th century physicist is not only a classic, but of great interest in the logic of science. Creating a new system of mechanics based upon space, time, and mass, it returns to axiomatic analysis, understanding of the formal or structural aspects of science, taking into account logic, observation, a priori elements. Of great historical importance to Poincaré, Carnap, Einstein, Milne. A 20 page introduction by R. S. Cohen, Wesleyan University, analyzes the implications of Hertz's thought and the logic of science. 13 page introduction by Helmholtz. xlii + 274pp. 5⅜ x 8. S316 Clothbound **$3.50**
S317 Paperbound **$1.75**

FROM MAGIC TO SCIENCE, Charles Singer. A great historian examines aspects of science from Roman Empire through Renaissance. Includes perhaps best discussion of early herbals, penetrating physiological interpretation of "The Visions of Hildegarde of Bingen." Also examines Arabian, Galenic influences; Pythagoras' sphere, Paracelsus; reawakening of science under Leonardo da Vinci, Vesalius; Lorica of Gildas the Briton; etc. Frequent quotations with translations from contemporary manuscripts. Unabridged, corrected edition. 158 unusual illustrations from Classical, Medieval sources. xxvii + 365pp. 5⅜ x 8. T390 Paperbound **$2.00**

A HISTORY OF THE CALCULUS, AND ITS CONCEPTUAL DEVELOPMENT, Carl B. Boyer. Provides laymen, mathematicians a detailed history of the development of the calculus, from beginnings in antiquity to final elaboration as mathematical abstraction. Gives a sense of mathematics not as technique, but as habit of mind, in progression of ideas of Zeno, Plato, Pythagoras, Eudoxus, Arabic and Scholastic mathematicians, Newton, Leibniz, Taylor, Descartes, Euler, Lagrange, Cantor, Weierstrass, and others. This first comprehensive, critical history of the calculus was originally entitled "The Concepts of the Calculus." Foreword by R. Courant. 22 figures. 25 page bibliography. v + 364pp. 5⅜ x 8.
S509 Paperbound **$2.00**

A DIDEROT PICTORIAL ENCYCLOPEDIA OF TRADES AND INDUSTRY, Manufacturing and the Technical Arts in Plates Selected from "L'Encyclopédie ou Dictionnaire Raisonné des Sciences, des Arts, et des Métiers" of Denis Diderot. Edited with text by C. Gillispie. First modern selection of plates from high-point of 18th century French engraving. Storehouse of technological information to historian of arts and science. Over 2,000 illustrations on 485 full page plates, most of them original size, show trades, industries of fascinating era in such great detail that modern reconstructions might be made of them. Plates teem with men, women, children performing thousands of operations; show sequence, general operations, closeups, details of machinery. Illustrates such important, interesting trades, industries as sowing, harvesting, beekeeping, tobacco processing, fishing, arts of war, mining, smelting, casting iron, extracting mercury, making gunpowder, cannons, bells, shoeing horses, tanning, papermaking, printing, dying, over 45 more categories. Professor Gillispie of Princeton supplies full commentary on all plates, identifies operations, tools, processes, etc. Material is presented in lively, lucid fashion. Of great interest to all studying history of science, technology. Heavy library cloth. 920pp. 9 x 12.
T421 2 volume set **$18.50**

DE MAGNETE, William Gilbert. Classic work on magnetism, founded new science. Gilbert was first to use word "electricity," to recognize mass as distinct from weight, to discover effect of heat on magnetic bodies; invented an electroscope, differentiated between static electricity and magnetism, conceived of earth as magnet. This lively work, by first great experimental scientist, is not only a valuable historical landmark, but a delightfully easy to follow record of a searching, ingenious mind. Translated by P. F. Mottelay. 25 page biographical memoir. 90 figures. lix + 368pp. 5⅜ x 8. S470 Paperbound **$2.00**

HISTORY OF MATHEMATICS, D. E. Smith. Most comprehensive, non-technical history of math in English. Discusses lives and works of over a thousand major, minor figures, with footnotes giving technical information outside book's scheme, and indicating disputed matters. Vol. I: A chronological examination, from primitive concepts through Egypt, Babylonia, Greece, the Orient, Rome, the Middle Ages, The Renaissance, and to 1900. Vol. II: The development of ideas in specific fields and problems, up through elementary calculus. "Marks an epoch . . . will modify the entire teaching of the history of science," George Sarton. 2 volumes, total of 510 illustrations, 1355pp. 5⅜ x 8. Set boxed in attractive container. T429, 430 Paperbound, the set **$5.00**

THE PHILOSOPHY OF SPACE AND TIME, H. Reichenbach. An important landmark in development of empiricist conception of geometry, covering foundations of geometry, time theory, consequences of Einstein's relativity, including: relations between theory and observations; coordinate definitions; relations between topological and metrical properties of space; psychological problem of visual intuition of non-Euclidean structures; many more topics important to modern science and philosophy. Majority of ideas require only knowledge of intermediate math. "Still the best book in the field," Rudolf Carnap. Introduction by R. Carnap. 49 figures. xviii + 296pp. 5⅜ x 8. S443 Paperbound **$2.00**

FOUNDATIONS OF SCIENCE: THE PHILOSOPHY OF THEORY AND EXPERIMENT, N. Campbell. A critique of the most fundamental concepts of science, particularly physics. Examines why certain propositions are accepted without question, demarcates science from philosophy, etc. Part I analyzes presuppositions of scientific thought: existence of material world, nature of laws, probability, etc; part 2 covers nature of experiment and applications of mathematics: conditions for measurement, relations between numerical laws and theories, error, etc. An appendix covers problems arising from relativity, force, motion, space, time. A classic in its field. "A real grasp of what science is," Higher Educational Journal. xiii + 565pp. 5⅝ x 8⅜. S372 Paperbound **$2.95**

THE STUDY OF THE HISTORY OF MATHEMATICS and **THE STUDY OF THE HISTORY OF SCIENCE, G. Sarton.** Excellent introductions, orientation, for beginning or mature worker. Describes duty of mathematical historian, incessant efforts and genius of previous generations. Explains how today's discipline differs from previous methods. 200 item bibliography with critical evaluations, best available biographies of modern mathematicians, best treatises on historical methods is especially valuable. 10 illustrations. 2 volumes bound as one. 113pp. + 75pp. 5⅜ x 8. T240 Paperbound **$1.25**

MATHEMATICAL PUZZLES

MATHEMATICAL PUZZLES OF SAM LOYD, selected and edited by **Martin Gardner.** 117 choice puzzles by greatest American puzzle creator and innovator, from his famous "Cyclopedia of Puzzles." All unique style, historical flavor of originals. Based on arithmetic, algebra, probability, game theory, route tracing, topology, sliding block, operations research, geometrical dissection. Includes famous "14-15" puzzle which was national craze, "Horse of a Different Color" which sold millions of copies. 120 line drawings, diagrams. Solutions. xx + 167pp. 5⅜ x 8. T498 Paperbound **$1.00**

SYMBOLIC LOGIC and THE GAME OF LOGIC, Lewis Carroll. "Symbolic Logic" is not concerned with modern symbolic logic, but is instead a collection of over 380 problems posed with charm and imagination, using the syllogism, and a fascinating diagrammatic method of drawing conclusions. In "The Game of Logic" Carroll's whimsical imagination devises a logical game played with 2 diagrams and counters (included) to manipulate hundreds of tricky syllogisms. The final section, "Hit or Miss" is a lagniappe of 101 additional puzzles in the delightful Carroll manner. Until this reprint edition, both of these books were rarities costing up to $15 each. Symbolic Logic: Index. xxxi + 199pp. The Game of Logic: 96pp. 2 vols. bound as one. 5⅜ x 8. T492 Paperbound **$1.50**

PILLOW PROBLEMS and A TANGLED TALE, Lewis Carroll. One of the rarest of all Carroll's works, "Pillow Problems" contains 72 original math puzzles, all typically ingenious. Particularly fascinating are Carroll's answers which remain exactly as he thought them out, reflecting his actual mental process. The problems in "A Tangled Tale" are in story form, originally appearing as a monthly magazine serial. Carroll not only gives the solutions, but uses answers sent in by readers to discuss wrong approaches and misleading paths, and grades them for insight. Both of these books were rarities until this edition, "Pillow Problems" costing up to $25, and "A Tangled Tale" $15. Pillow Problems: Preface and Introduction by Lewis Carroll. xx + 109pp. A Tangled Tale: 6 illustrations. 152pp. Two vols. bound as one. 5⅜ x 8. T493 Paperbound **$1.50**

NEW WORD PUZZLES, G. L. Kaufman. 100 brand new challenging puzzles on words, combinations, never before published. Most are new types invented by author, for beginners and experts both. Squares of letters follow chess moves to build words; symmetrical designs made of synonyms; rhymed crostics; double word squares; syllable puzzles where you fill in missing syllables instead of missing letter; many other types, all new. Solutions. "Excellent," Recreation. 100 puzzles. 196 figures. vi + 122pp. 5⅜ x 8. T344 Paperbound **$1.00**

MATHEMATICAL EXCURSIONS, H. A. Merrill. Fun, recreation, insights into elementary problem solving. Math expert guides you on by-paths not generally travelled in elementary math courses—divide by inspection, Russian peasant multiplication; memory systems for pi; odd, even magic squares; dyadic systems; square roots by geometry; Tchebichev's machine; dozens more. Solutions to more difficult ones. "Brain stirring stuff . . . a classic," Genie. 50 illustrations. 145pp. 5⅜ x 8. T350 Paperbound **$1.00**

THE BOOK OF MODERN PUZZLES, G. L. Kaufman. Over 150 puzzles, absolutely all new material based on same appeal as crosswords, deduction puzzles, but with different principles, techniques. 2-minute teasers, word labyrinths, design, pattern, logic, observation puzzles, puzzles testing ability to apply general knowledge to peculiar situations, many others. Solutions. 116 illustrations. 192pp. 5⅜ x 8. T143 Paperbound **$1.00**

MATHEMAGIC, MAGIC PUZZLES, AND GAMES WITH NUMBERS, R. V. Heath. Over 60 puzzles, stunts, on properties of numbers. Easy techniques for multiplying large numbers mentally, identifying unknown numbers, finding date of any day in any year. Includes The Lost Digit, 3 Acrobats, Psychic Bridge, magic squares, triangles, cubes, others not easily found elsewhere. Edited by J. S. Meyer. 76 illustrations. 128pp. 5⅜ x 8. T110 Paperbound **$1.00**

DOVER SCIENCE BOOKS

PUZZLE QUIZ AND STUNT FUN, J. Meyer. 238 high-priority puzzles, stunts, tricks—math puzzles like The Clever Carpenter, Atom Bomb, Please Help Alice; mysteries, deductions like The Bridge of Sighs, Secret Code; observation puzzlers like The American Flag, Playing Cards, Telephone Dial; over 200 others with magic squares, tongue twisters, puns, anagrams. Solutions. Revised, enlarged edition of "Fun-To-Do." Over 100 illustrations. 238 puzzles, stunts, tricks. 256pp. 5⅜ x 8. T337 Paperbound **$1.00**

101 PUZZLES IN THOUGHT AND LOGIC, C. R. Wylie, Jr. For readers who enjoy challenge, stimulation of logical puzzles without specialized math or scientific knowledge. Problems entirely new, range from relatively easy to brainteasers for hours of subtle entertainment. Detective puzzles, find the lying fisherman, how a blind man identifies color by logic, many more. Easy-to-understand introduction to logic of puzzle solving and general scientific method. 128pp. 5⅜ x 8. T367 Paperbound **$1.00**

CRYPTANALYSIS, H. F. Gaines. Standard elementary, intermediate text for serious students. Not just old material, but much not generally known, except to experts. Concealment, Transposition, Substitution ciphers; Vigenere, Kasiski, Playfair, multafid, dozens of other techniques. Formerly "Elementary Cryptanalysis." Appendix with sequence charts, letter frequencies in English, 5 other languages, English word frequencies. Bibliography. 167 codes. New to this edition: solutions to codes. vi + 230pp. 5⅜ x 8⅜. T97 Paperbound **$1.95**

CRYPTOGRAPY, L. D. Smith. Excellent elementary introduction to enciphering, deciphering secret writing. Explains transposition, substitution ciphers; codes; solutions; geometrical patterns, route transcription, columnar transposition, other methods. Mixed cipher systems; single, polyalphabetical substitutions; mechanical devices; Vigenere; etc. Enciphering Japanese; explanation of Baconian biliteral cipher; frequency tables. Over 150 problems. Bibliography. Index. 164pp. 5⅜ x 8. T247 Paperbound **$1.00**

MATHEMATICS, MAGIC AND MYSTERY, M. Gardner. Card tricks, metal mathematics, stage mind-reading, other "magic" explained as applications of probability, sets, number theory, etc. Creative examination of laws, applications. Scores of new tricks, insights. 115 sections on cards, dice, coins; vanishing tricks, many others. No sleight of hand—math guarantees success. "Could hardly get more entertainment . . . easy to follow," Mathematics Teacher. 115 illustrations. xii + 174pp. 5⅜ x 8. T335 Paperbound **$1.00**

AMUSEMENTS IN MATHEMATICS, H. E. Dudeney. Foremost British originator of math puzzles, always witty, intriguing, paradoxical in this classic. One of largest collections. More than 430 puzzles, problems, paradoxes. Mazes, games, problems on number manipulations, unicursal, other route problems, puzzles on measuring, weighing, packing, age, kinship, chessboards, joiners', crossing river, plane figure dissection, many others. Solutions. More than 450 illustrations. viii + 258pp. 5⅜ x 8. T473 Paperbound **$1.25**

THE CANTERBURY PUZZLES H. E. Dudeney. Chaucer's pilgrims set one another problems in story form. Also Adventures of the Puzzle Club, the Strange Escape of the King's Jester, the Monks of Riddlewell, the Squire's Christmas Puzzle Party, others. All puzzles are original, based on dissecting plane figures, arithmetic, algebra, elementary calculus, other branches of mathematics, and purely logical ingenuity. "The limit of ingenuity and intricacy," The Observer. Over 110 puzzles, full solutions. 150 illustrations. viii + 225 pp. 5⅜ x 8. T474 Paperbound **$1.25**

MATHEMATICAL PUZZLES FOR BEGINNERS AND ENTHUSIASTS, G. Mott-Smith. 188 puzzles to test mental agility. Inference, interpretation, algebra, dissection of plane figures, geometry, properties of numbers, decimation, permutations, probability, all are in these delightful problems. Includes the Odic Force, How to Draw an Ellipse, Spider's Cousin, more than 180 others. Detailed solutions. Appendix with square roots, triangular numbers, primes, etc. 135 illustrations. 2nd revised edition. 248pp. 5⅜ x 8. T198 Paperbound **$1.00**

MATHEMATICAL RECREATIONS, M. Kraitchik. Some 250 puzzles, problems, demonstrations of recreation mathematics on relatively advanced level. Unusual historical problems from Greek, Medieval, Arabic, Hindu sources; modern problems on "mathematics without numbers," geometry, topology, arithmetic, etc. Pastimes derived from figurative, Mersenne, Fermat numbers: fairy chess; latruncles: reversi; etc. Full solutions. Excellent insights into special fields of math. "Strongly recommended to all who are interested in the lighter side of mathematics," Mathematical Gaz. 181 illustrations. 330pp. 5⅜ x 8. T163 Paperbound **$1.75**

FICTION

FLATLAND, E. A. Abbott. A perennially popular science-fiction classic about life in a 2-dimensional world, and the impingement of higher dimensions. Political, satiric, humorous, moral overtones. This land where women are straight lines and the lowest and most dangerous classes are isosceles triangles with 3° vertices conveys brilliantly a feeling for many concepts of modern science. 7th edition. New introduction by Banesh Hoffmann. 128pp. 5⅜ x 8. T1 Paperbound **$1.00**

SEVEN SCIENCE FICTION NOVELS OF H. G. WELLS. Complete texts, unabridged, of seven of Wells' greatest novels: The War of the Worlds, The Invisible Man, The Island of Dr. Moreau, The Food of the Gods, First Men in the Moon, In the Days of the Comet, The Time Machine. Still considered by many experts to be the best science-fiction ever written, they will offer amusements and instruction to the scientific minded reader. "The great master," Sky and Telescope. 1051pp. 5⅜ x 8. T264 Clothbound **$3.95**

28 SCIENCE FICTION STORIES OF H. G. WELLS. Unabridged! This enormous omnibus contains 2 full length novels—Men Like Gods, Star Begotten—plus 26 short stories of space, time, invention, biology, etc. The Crystal Egg, The Country of the Blind, Empire of the Ants, The Man Who Could Work Miracles, Aepyornis Island, A Story of the Days to Come, and 20 others "A master . . . not surpassed by . . . writers of today," The English Journal. 915pp. 5⅜ x 8. T265 Clothbound **$3.95**

FIVE ADVENTURE NOVELS OF H. RIDER HAGGARD. All the mystery and adventure of darkest Africa captured accurately by a man who lived among Zulus for years, who knew African ethnology, folkways as did few of his contemporaries. They have been regarded as examples of the very best high adventure by such critics as Orwell, Andrew Lang, Kipling. Contents: She, King Solomon's Mines, Allan Quatermain, Allan's Wife, Maiwa's Revenge. "Could spin a yarn so full of suspense and color that you couldn't put the story down," Sat. Review. 821pp. 5⅜ x 8. T108 Clothbound **$3.95**

CHESS AND CHECKERS

LEARN CHESS FROM THE MASTERS, Fred Reinfeld. Easiest, most instructive way to improve your game—play 10 games against such masters as Marshall, Znosko-Borovsky, Bronstein, Najdorf, etc., with each move graded by easy system. Includes ratings for alternate moves possible. Games selected for interest, clarity, easily isolated principles. Covers Ruy Lopez, Dutch Defense, Vienna Game openings; subtle, intricate middle game variations; all-important end game. Full annotations. Formerly "Chess by Yourself." 91 diagrams. viii + 144pp. 5⅜ x 8. T362 Paperbound **$1.00**

REINFELD ON THE END GAME IN CHESS, Fred Reinfeld. Analyzes 62 end games by Alekhine, Flohr, Tarrasch, Morphy, Capablanca, Rubinstein, Lasker, Reshevsky, other masters. Only 1st rate book with extensive coverage of error—tell exactly what is wrong with each move you might have made. Centers around transitions from middle play to end play. King and pawn, minor pieces, queen endings; blockage, weak, passed pawns, etc. "Excellent . . . a boon," Chess Life. Formerly "Practical End Play." 62 figures. vi + 177pp. 5⅜ x 8. **T417 Paperbound $1.25**

HYPERMODERN CHESS as developed in the games of its greatest exponent, ARON NIMZOVICH, edited by Fred Reinfeld. An intensely original player, analyst, Nimzovich's approaches startled, often angered the chess world. This volume, designed for the average player, shows how his iconoclastic methods won him victories over Alekhine, Lasker, Marshall, Rubinstein, Spielmann, others, and infused new life into the game. Use his methods to startle opponents, invigorate play. "Annotations and introductions to each game . . . are excellent," Times (London). 180 diagrams. viii + 220pp. 5⅜ x 8. T448 Paperbound **$1.35**

THE ADVENTURE OF CHESS, Edward Lasker. Lively reader, by one of America's finest chess masters, including: history of chess, from ancient Indian 4-handed game of Chaturanga to great players of today; such delights and oddities as Maelzel's chess-playing automaton that beat Napoleon 3 times; etc. One of most valuable features is author's personal recollections of men he has played against—Nimzovich, Emanuel Lasker, Capablanca, Alekhine, etc. Discussion of chess-playing machines (newly revised). 5 page chess primer. 11 illustrations. 53 diagrams. 296pp. 5⅜ x 8. S510 Paperbound **$1.45**

THE ART OF CHESS, James Mason. Unabridged reprinting of latest revised edition of most famous general study ever written. Mason, early 20th century master, teaches beginning, intermediate player over 90 openings; middle game, end game, to see more moves ahead, to plan purposefully, attack, sacrifice, defend, exchange, govern general strategy. "Classic . . . one of the clearest and best developed studies," Publishers Weekly. Also included, a complete supplement by F. Reinfeld, "How Do You Play Chess?", invaluable to beginners for its lively question-and-answer method. 448 diagrams. 1947 Reinfeld-Bernstein text. Bibliography. xvi + 340pp. 5⅜ x 8. T463 Paperbound **$1.85**

MORPHY'S GAMES OF CHESS, edited by P. W. Sergeant. Put boldness into your game by flowing brilliant, forceful moves of the greatest chess player of all time. 300 of Morphy's best games, carefully annotated to reveal principles. 54 classics against masters like Anderssen, Harrwitz, Bird, Paulsen, and others. 52 games at odds; 54 blindfold games; plus over 100 others. Follow his interpretation of Dutch Defense, Evans Gambit, Giuoco Piano, Ruy Lopez, many more. Unabridged reissue of latest revised edition. New introduction by F. Reinfeld. Annotations, introduction by Sergeant. 235 diagrams. x + 352pp. 5⅜ x 8. T386 Paperbound **$1.75**

WIN AT CHECKERS, M. Hopper. (Formerly "Checkers.") Former World's Unrestricted Checker Champion discusses principles of game, expert's shots, traps, problems for beginner, standard openings, locating best move, end game, opening "blitzkrieg" moves to draw when behind, etc. Over 100 detailed questions, answers anticipate problems. Appendix. 75 problems with solutions, diagrams. 79 figures. xi + 107pp. 5⅜ x 8. T363 Paperbound **$1.00**

HOW TO FORCE CHECKMATE, Fred Reinfeld. If you have trouble finishing off your opponent, here is a collection of lightning strokes and combinations from actual tournament play. Starts with 1-move checkmates, works up to 3-move mates. Develops ability to lock ahead, gain new insights into combinations, complex or deceptive positions; ways to estimate weaknesses, strengths of you and your opponent. "A good deal of amusement and instruction," Times, (London). 300 diagrams. Solutions to all positions. Formerly "Challenge to Chess Players." 111pp. 5⅜ x 8. T417 Paperbound **$1.25**

A TREASURY OF CHESS LORE, edited by Fred Reinfeld. Delightful collection of anecdotes, short stories, aphorisms by, about masters; poems, accounts of games, tournaments, photographs; hundreds of humorous, pithy, satirical, wise, historical episodes, comments, word portraits. Fascinating "must" for chess players; revealing and perhaps seductive to those who wonder what their friends see in game. 49 photographs (14 full page plates). 12 diagrams. xi + 306pp. 5⅜ x 8. T458 Paperbound **$1.75**

WIN AT CHESS, Fred Reinfeld. 300 practical chess situations, to sharpen your eye, test skill against masters. Start with simple examples, progress at own pace to complexities. This selected series of crucial moments in chess will stimulate imagination, develop stronger, more versatile game. Simple grading system enables you to judge progress. "Extensive use of diagrams is a great attraction," Chess. 300 diagrams. Notes, solutions to every situation. Formerly "Chess Quiz." vi + 120pp. 5⅜ x 8. T433 Paperbound **$1.00**

MATHEMATICS: ELEMENTARY TO INTERMEDIATE

HOW TO CALCULATE QUICKLY, H. Sticker. Tried and true method to help mathematics of everyday life. Awakens "number sense"—ability to see relationships between numbers as whole quantities. A serious course of over 9000 problems and their solutions through techniques not taught in schools: left-to-right multiplications, new fast division, etc. 10 minutes a day will double or triple calculation speed. Excellent for scientist at home in higher math, but dissatisfied with speed and accuracy in lower math. 256pp. 5 x 7¼. Paperbound **$1.00**

FAMOUS PROBLEMS OF ELEMENTARY GEOMETRY, Felix Klein. Expanded version of 1894 Easter lectures at Göttingen. 3 problems of classical geometry: squaring the circle, trisecting angle, doubling cube, considered with full modern implications: transcendental numbers, pi, etc. "A modern classic . . . no knowledge of higher mathematics is required," Scientia. Notes by R. Archibald. 16 figures. xi + 92pp. 5⅜ x 8. T298 Paperbound **$1.00**

HIGHER MATHEMATICS FOR STUDENTS OF CHEMISTRY AND PHYSICS, J. W. Mellor. Practical, not abstract, building problems out of familiar laboratory material. Covers differential calculus, coordinate, analytical geometry, functions, integral calculus, infinite series, numerical equations, differential equations, Fourier's theorem probability, theory of errors, calculus of variations, determinants. "If the reader is not familiar with this book, it will repay him to examine it," Chem. and Engineering News. 800 problems. 189 figures. xxi + 641pp. 5⅜ x 8. S193 Paperbound **$2.25**

TRIGONOMETRY REFRESHER FOR TECHNICAL MEN, A. A. Klaf. 913 detailed questions, answers cover most important aspects of plane, spherical trigonometry—particularly useful in clearing up difficulties in special areas. Part I: plane trig, angles, quadrants, functions, graphical representation, interpolation, equations, logs, solution of triangle, use of slide rule, etc. Next 188 pages discuss applications to navigation, surveying, elasticity, architecture, other special fields. Part 3: spherical trig, applications to terrestrial, astronomical problems. Methods of time-saving, simplification of principal angles, make book most useful. 913 questions answered. 1738 problems, answers to odd numbers. 494 figures. 24 pages of formulas, functions. x + 629pp. 5⅜ x 8. T371 Paperbound **$2.00**

CALCULUS REFRESHER FOR TECHNICAL MEN, A. A. Klaf. 756 questions examine most important aspects of integral, differential calculus. Part I: simple differential calculus, constants, variables, functions, increments, logs, curves, etc. Part 2: fundamental ideas of integrations, inspection, substitution, areas, volumes, mean value, double, triple integration, etc. Practical aspects stressed. 50 pages illustrate applications to specific problems of civil, nautical engineering, electricity, stress, strain, elasticity, similar fields. 756 questions answered. 566 problems, mostly answered. 36pp. of useful constants, formulas. v + 431pp. 5⅜ x 8. T370 Paperbound **$2.00**

MONOGRAPHS ON TOPICS OF MODERN MATHEMATICS, edited by J. W. A. Young. Advanced mathematics for persons who have forgotten, or not gone beyond, high school algebra. 9 monographs on foundation of geometry, modern pure geometry, non-Euclidean geometry, fundamental propositions of algebra, algebraic equations, functions, calculus, theory of numbers, etc. Each monograph gives proofs of important results, and descriptions of leading methods, to provide wide coverage. "Of high merit," Scientific American. New introduction by Prof. M. Kline, N.Y. Univ. 100 diagrams. xvi + 416pp. 6⅛ x 9¼.
S289 Paperbound **$2.00**

MATHEMATICS IN ACTION, O. G. Sutton. Excellent middle level application of mathematics to study of universe, demonstrates how math is applied to ballistics, theory of computing machines, waves, wave-like phenomena, theory of fluid flow, meteorological problems, statistics, flight, similar phenomena. No knowledge of advanced math required. Differential equations, Fourier series, group concepts, Eigenfunctions, Planck's constant, airfoil theory, and similar topics explained so clearly in everyday language that almost anyone can derive benefit from reading this even if much of high-school math is forgotten. 2nd edition. 88 figures. viii + 236pp. 5⅜ x 8. T450 Clothbound **$3.50**

ELEMENTARY MATHEMATICS FROM AN ADVANCED STANDPOINT, Felix Klein. Classic text, an outgrowth of Klein's famous integration and survey course at Göttingen. Using one field to interpret, adjust another, it covers basic topics in each area, with extensive analysis. Especially valuable in areas of modern mathematics. "A great mathematician, inspiring teacher, . . . deep insight," Bul., Amer. Math Soc.

Vol. I. ARITHMETIC, ALGEBRA, ANALYSIS. Introduces concept of function immediately, enlivens discussion with graphical, geometric methods. Partial contents: natural numbers, special properties, complex numbers. Real equations with real unknowns, complex quantities. Logarithmic, exponential functions, infinitesimal calculus. Transcendence of e and pi, theory of assemblages. Index. 125 figures. ix + 274pp. 5⅜ x 8. S151 Paperbound **$1.75**

Vol. II. GEOMETRY. Comprehensive view, accompanies space perception inherent in geometry with analytic formulas which facilitate precise formulation. Partial contents: Simplest geometric manifold; line segments, Grassman determinant principles, classication of configurations of space. Geometric transformations: affine, projective, higher point transformations, theory of the imaginary. Systematic discussion of geometry and its foundations. 141 illustrations. ix + 214pp. 5⅜ x 8. S151 Paperbound **$1.75**

A TREATISE ON PLANE AND ADVANCED TRIGONOMETRY, E. W. Hobson. Extraordinarily wide coverage, going beyond usual college level, one of few works covering advanced trig in full detail. By a great expositor with unerring anticipation of potentially difficult points. Includes circular functions; expansion of functions of multiple angle; trig tables; relations between sides, angles of triangles; complex numbers; etc. Many problems fully solved. "The best work on the subject," Nature. Formerly entitled "A Treatise on Plane Trigonometry." 689 examples. 66 figures. xvi + 383pp. 5⅜ x 8. S353 Paperbound **$1.95**

NON-EUCLIDEAN GEOMETRY, Roberto Bonola. The standard coverage of non-Euclidean geometry. Examines from both a historical and mathematical point of view geometries which have arisen from a study of Euclid's 5th postulate on parallel lines. Also included are complete texts, translated, of Bolyai's "Theory of Absolute Space," Lobachevsky's "Theory of Parallels." 180 diagrams. 431pp. 5⅜ x 8. S27 Paperbound **$1.95**

GEOMETRY OF FOUR DIMENSIONS, H. P. Manning. Unique in English as a clear, concise introduction. Treatment is synthetic, mostly Euclidean, though in hyperplanes and hyperspheres at infinity, non-Euclidean geometry is used. Historical introduction. Foundations of 4-dimensional geometry. Perpendicularity, simple angles. Angles of planes, higher order. Symmetry, order, motion; hyperpyramids, hypercones, hyperspheres; figures with parallel elements; volume, hypervolume in space; regular polyhedroids. Glossary. 78 figures. ix + 348pp. 5⅜ x 8. S182 Paperbound **$1.95**

MATHEMATICS: INTERMEDIATE TO ADVANCED

GEOMETRY (EUCLIDEAN AND NON-EUCLIDEAN)

THE GEOMETRY OF RENÉ DESCARTES. With this book, Descartes founded analytical geometry. Original French text, with Descartes's own diagrams, and excellent Smith-Latham translation. Contains: Problems the Construction of Which Requires only Straight Lines and Circles; On the Nature of Curved Lines; On the Construction of Solid or Supersolid Problems. Diagrams. 258pp. 5⅜ x 8. S68 Paperbound **$1.50**

THE WORKS OF ARCHIMEDES, edited by T. L. Heath. All the known works of the great Greek mathematician, including the recently discovered Method of Archimedes. Contains: On Sphere and Cylinder, Measurement of a Circle, Spirals, Conoids, Spheroids, etc. Definitive edition of greatest mathematical intellect of ancient world. 186 page study by Heath discusses Archimedes and history of Greek mathematics. 563pp. 5⅜ x 8. S9 Paperbound **$2.00**

COLLECTED WORKS OF BERNARD RIEMANN. Important sourcebook, first to contain complete text of 1892 "Werke" and the 1902 supplement, unabridged. 31 monographs, 3 complete lecture courses, 15 miscellaneous papers which have been of enormous importance in relativity, topology, theory of complex variables, other areas of mathematics. Edited by R. Dedekind, H. Weber, M. Noether, W. Wirtinger. German text; English introduction by Hans Lewy. 690pp. 5⅜ x 8. S226 Paperbound **$2.85**

THE THIRTEEN BOOKS OF EUCLID'S ELEMENTS, edited by Sir Thomas Heath. Definitive edition of one of very greatest classics of Western world. Complete translation of Heiberg text, plus spurious Book XIV. 150 page introduction on Greek, Medieval mathematics, Euclid, texts, commentators, etc. Elaborate critical apparatus parallels text, analyzing each definition, postulate, proposition, covering textual matters, refutations, supports, extrapolations, etc. This is the full Euclid. Unabridged reproduction of Cambridge U. 2nd edition. 3 volumes. 995 figures. 1426pp. 5⅜ x 8. S88, 89, 90, 3 volume set, paperbound **$6.00**

AN INTRODUCTION TO GEOMETRY OF N DIMENSIONS, D. M. Y. Sommerville. Presupposes no previous knowledge of field. Only book in English devoted exclusively to higher dimensional geometry. Discusses fundamental ideas of incidence, parallelism, perpendicularity, angles between linear space, enumerative geometry, analytical geometry from projective and metric views, polytopes, elementary ideas in analysis situs, content of hyperspacial figures. 60 diagrams. 196pp. 5⅜ x 8. S494 Paperbound **$1.50**

ELEMENTS OF NON-EUCLIDEAN GEOMETRY, D. M. Y. Sommerville. Unique in proceeding step-by-step. Requires only good knowledge of high-school geometry and algebra, to grasp elementary hyperbolic, elliptic, analytic non-Euclidean Geometries; space curvature and its implications; radical axes; homopethic centres and systems of circles; parataxy and parallelism; Gauss' proof of defect area theorem; much more, with exceptional clarity. 126 problems at chapter ends. 133 figures. xvi + 274pp. 5⅜ x 8. S460 Paperbound **$1.50**

THE FOUNDATIONS OF EUCLIDEAN GEOMETRY, H. G. Forder. First connected, rigorous account in light of modern analysis, establishing propositions without recourse to empiricism, without multiplying hypotheses. Based on tools of 19th and 20th century mathematicians, who made it possible to remedy gaps and complexities, recognize problems not earlier discerned. Begins with important relationship of number systems in geometrical figures. Considers classes, relations, linear order, natural numbers, axioms for magnitudes, groups, quasi-fields, fields, non-Archimedian systems, the axiom system (at length), particular axioms (two chapters on the Parallel Axioms), constructions, congruence, similarity, etc. Lists: axioms employed, constructions, symbols in frequent use. 295pp. 5⅜ x 8. S481 Paperbound **$2.00**

CALCULUS, FUNCTION THEORY (REAL AND COMPLEX), FOURIER THEORY

FIVE VOLUME "THEORY OF FUNCTIONS" SET BY KONRAD KNOPP. Provides complete, readily followed account of theory of functions. Proofs given concisely, yet without sacrifice of completeness or rigor. These volumes used as texts by such universities as M.I.T., Chicago, N.Y. City College, many others. "Excellent introduction . . . remarkably readable, concise, clear, rigorous," J. of the American Statistical Association.

ELEMENTS OF THE THEORY OF FUNCTIONS, Konrad Knopp. Provides background for further volumes in this set, or texts on similar level. Partial contents: Foundations, system of complex numbers and Gaussian plane of numbers, Riemann sphere of numbers, mapping by linear functions, normal forms, the logarithm, cyclometric functions, binomial series. "Not only for the young student, but also for the student who knows all about what is in it," Mathematical Journal. 140pp. 5⅜ x 8. S154 Paperbound **$1.35**

THEORY OF FUNCTIONS, PART I, Konrad Knopp. With volume II, provides coverage of basic concepts and theorems. Partial contents: numbers and points, functions of a complex variable, integral of a continuous function, Cauchy's intergral theorem, Cauchy's integral formulae, series with variable terms, expansion and analytic function in a power series, analytic continuation and complete definition of analytic functions, Laurent expansion, types of singularities. vii + 146pp. 5⅜ x 8. S156 Paperbound **$1.35**

THEORY OF FUNCTIONS, PART II, Konrad Knopp. Application and further development of general theory, special topics. Single valued functions, entire, Weierstrass. Meromorphic functions: Mittag-Leffler. Periodic functions. Multiple valued functions. Riemann surfaces. Algebraic functions. Analytical configurations, Riemann surface. x + 150pp. 5⅜ x 8. S157 Paperbound **$1.35**

PROBLEM BOOK IN THE THEORY OF FUNCTIONS, VOLUME I, Konrad Knopp. Problems in elementary theory, for use with Knopp's "Theory of Functions," or any other text. Arranged according to increasing difficulty. Fundamental concepts, sequences of numbers and infinite series, complex variable, integral theorems, development in series, conformal mapping. Answers. viii + 126pp. 5⅜ x 8. S 158 **Paperbound $1.35**

PROBLEM BOOK IN THE THEORY OF FUNCTIONS, VOLUME II, Konrad Knopp. Advanced theory of functions, to be used with Knopp's "Theory of Functions," or comparable text. Singularities, entire and meromorphic functions, periodic, analytic, continuation, multiple-valued functions, Riemann surfaces, conformal mapping. Includes section of elementary problems. "The difficult task of selecting . . . problems just within the reach of the beginner is here masterfully accomplished," AM. MATH. SOC. Answers. 138pp. 5⅜ x 8. S159 Paperbound **$1.35**

ADVANCED CALCULUS, E. B. Wilson. Still recognized as one of most comprehensive, useful texts. Immense amount of well-represented, fundamental material, including chapters on vector functions, ordinary differential equations, special functions, calculus of variations, etc., which are excellent introductions to these areas. Requires only one year of calculus. Over 1300 exercises cover both pure math and applications to engineering and physical problems. Ideal reference, refresher. 54 page introductory review. ix + 566pp. 5⅜ x 8. S504 Paperbound **$2.45**

LECTURES ON THE THEORY OF ELLIPTIC FUNCTIONS, H. Hancock. Reissue of only book in English with so extensive a coverage, especially of Abel, Jacobi, Legendre, Weierstrass, Hermite, Liouville, and Riemann. Unusual fullness of treatment, plus applications as well as theory in discussing universe of elliptic integrals, originating in works of Abel and Jacobi. Use is made of Riemann to provide most general theory. 40-page table of formulas. 76 figures. xxiii + 498pp. 5⅜ x 8. S483 Paperbound **$2.55**

THEORY OF FUNCTIONALS AND OF INTEGRAL AND INTEGRO-DIFFERENTIAL EQUATIONS, Vito Volterra. Unabridged republication of only English translation, General theory of functions depending on continuous set of values of another function. Based on author's concept of transition from finite number of variables to a continually infinite number. Includes much material on calculus of variations. Begins with fundamentals, examines generalization of analytic functions, functional derivative equations, applications, other directions of theory, etc. New introduction by G. C. Evans. Biography, criticism of Volterra's work by E. Whittaker. xxxx + 226pp. 5⅜ x 8. S502 Paperbound **$1.75**

AN INTRODUCTION TO FOURIER METHODS AND THE LAPLACE TRANSFORMATION, Philip Franklin. Concentrates on essentials, gives broad view, suitable for most applications. Requires only knowledge of calculus. Covers complex qualities with methods of computing elementary functions for complex values of argument and finding approximations by charts; Fourier series; harmonic anaylsis; much more. Methods are related to physical problems of heat flow, vibrations, electrical transmission, electromagnetic radiation, etc. 828 problems, answers. Formerly entitled "Fourier Methods." x + 289pp. 5⅜ x 8. S452 Paperbound **$1.75**

THE ANALYTICAL THEORY OF HEAT, Joseph Fourier. This book, which revolutionized mathematical physics, has been used by generations of mathematicians and physicists interested in heat or application of Fourier integral. Covers cause and reflection of rays of heat, radiant heating, heating of closed spaces, use of trigonometric series in theory of heat, Fourier integral, etc. Translated by Alexander Freeman. 20 figures. xxii + 466pp. 5⅜ x 8. S93 Paperbound **$2.00**

ELLIPTIC INTEGRALS, H. Hancock. Invaluable in work involving differential equations with cubics, quatrics under root sign, where elementary calculus methods are inadequate. Practical solutions to problems in mathematics, engineering, physics; differential equations requiring integration of Lamé's, Briot's, or Bouquet's equations; determination of arc of ellipse, hyperbola, lemiscate; solutions of problems in elastics; motion of a projectile under resistance varying as the cube of the velocity; pendulums; more. Exposition in accordance with Legendre-Jacobi theory. Rigorous discussion of Legendre transformations. 20 figures. 5 place table. 104pp. 5⅜ x 8. S484 Paperbound **$1.25**

THE TAYLOR SERIES, AN INTRODUCTION TO THE THEORY OF FUNCTIONS OF A COMPLEX VARIABLE, P. Dienes. Uses Taylor series to approach theory of functions, using ordinary calculus only, except in last 2 chapters. Starts with introduction to real variable and complex algebra, derives properties of infinite series, complex differentiation, integration, etc. Covers biuniform mapping, overconvergence and gap theorems, Taylor series on its circle of convergence, etc. Unabridged corrected reissue of first edition. 186 examples, many fully worked out. 67 figures. xii + 555pp. 5⅜ x 8. S391 Paperbound **$2.75**

LINEAR INTEGRAL EQUATIONS, W. V. Lovitt. Systematic survey of general theory, with some application to differential equations, calculus of variations, problems of math, physics. Includes: integral equation of 2nd kind by successive substitutions; Fredholm's equation as ratio of 2 integral series in lambda, applications of the Fredholm theory, Hilbert-Schmidt theory of symmetric kernels, application, etc. Neumann, Dirichlet, vibratory problems. ix + 253pp. 5⅜ x 8. S175 Clothbound **$3.50**
S176 Paperbound **$1.60**

DICTIONARY OF CONFORMAL REPRESENTATIONS, H. Kober. Developed by British Admiralty to solve Laplace's equation in 2 dimensions. Scores of geometrical forms and transformations for electrical engineers, Joukowski aerofoil for aerodynamics, Schwartz-Christoffel transformations for hydro-dynamics, transcendental functions. Contents classified according to analytical functions describing transformations with corresponding regions. Glossary. Topological index. 447 diagrams. 6⅛ x 9¼. S160 Paperbound **$2.00**

ELEMENTS OF THE THEORY OF REAL FUNCTIONS, J. E. Littlewood. Based on lectures at Trinity College, Cambridge, this book has proved extremely successful in introducing graduate students to modern theory of functions. Offers full and concise coverage of classes and cardinal numbers, well ordered series, other types of series, and elements of the theory of sets of points. 3rd revised edition. vii + 71pp. 5⅜ x 8. S171 Clothbound **$2.85**
S172 Paperbound **$1.25**

INFINITE SEQUENCES AND SERIES, Konrad Knopp. 1st publication in any language. Excellent introduction to 2 topics of modern mathematics, designed to give student background to penetrate further alone. Sequences and sets, real and complex numbers, etc. Functions of a real and complex variable. Sequences and series. Infinite series. Convergent power series. Expansion of elementary functions. Numerical evaluation of series. v + 186pp. 5⅜ x 8.
S152 Clothbound **$3.50**
S153 Paperbound **$1.75**

THE THEORY AND FUNCTIONS OF A REAL VARIABLE AND THE THEORY OF FOURIER'S SERIES, E. W .Hobson. One of the best introductions to set theory and various aspects of functions and Fourier's series. Requires only a good background in calculus. Exhaustive coverage of: metric and descriptive properties of sets of points; transfinite numbers and order types; functions of a real variable; the Riemann and Lebesgue integrals; sequences and series of numbers; power-series; functions representable by series sequences of continuous functions; trigonometrical series; representation of functions by Fourier's series; and much more. "The best possible guide," Nature. Vol. I: 88 detailed examples, 10 figures. Index. xv + 736pp. Vol. II: 117 detailed examples, 13 figures. x + 780pp. 6⅛ x 9¼.
Vol. I: S387 Paperbound **$3.00**
Vol. II: S388 Paperbound **$3.00**

ALMOST PERIODIC FUNCTIONS, A. S. Besicovitch. Unique and important summary by a well known mathematician covers in detail the two stages of development in Bohr's theory of almost periodic functions: (1) as a generalization of pure periodicity, with results and proofs; (2) the work done by Stepanof, Wiener, Weyl, and Bohr in generalizing the theory. xi + 180pp. 5⅜ x 8. S18 Paperbound **$1.75**

INTRODUCTION TO THE THEORY OF FOURIER'S SERIES AND INTEGRALS, H. S. Carslaw. 3rd revised edition, an outgrowth of author's courses at Cambridge. Historical introduction, rational, irrational numbers, infinite sequences and series, functions of a single variable, definite integral, Fourier series, and similar topics. Appendices discuss practical harmonic analysis, periodogram analysis, Lebesgue's theory. 84 examples. xiii + 368pp. 5⅜ x 8.
S48 Paperbound **$2.00**

SYMBOLIC LOGIC

THE ELEMENTS OF MATHEMATICAL LOGIC, Paul Rosenbloom. First publication in any language. For mathematically mature readers with no training in symbolic logic. Development of lectures given at Lund Univ., Sweden, 1948. Partial contents: Logic of classes, fundamental theorems, Boolean algebra, logic of propositions, of propositional functions, expressive languages, combinatory logics, development of math within an object language, paradoxes, theorems of Post, Goedel, Church, and similar topics. iv + 214pp. 5⅜ x 8.
S227 Paperbound **$1.45**

INTRODUCTION TO SYMBOLIC LOGIC AND ITS APPLICATION, R. Carnap. Clear, comprehensive, rigorous, by perhaps greatest living master. Symbolic languages analyzed, one constructed. Applications to math (axiom systems for set theory, real, natural numbers), topology (Dedekind, Cantor continuity explanations), physics (general analysis of determination, causality, space-time topology), biology (axiom system for basic concepts). "A masterpiece," Zentralblatt für Mathematik und Ihre Grenzgebiete. Over 300 exercises. 5 figures. xvi + 241pp. 5⅜ x 8. S453 Paperbound **$1.85**

AN INTRODUCTION TO SYMBOLIC LOGIC, Susanne K. Langer. Probably clearest book for the philosopher, scientist, layman—no special knowledge of math required. Starts with simplest symbols, goes on to give remarkable grasp of Boole-Schroeder, Russell-Whitehead systems, clearly, quickly. Partial Contents: Forms, Generalization, Classes, Deductive System of Classes, Algebra of Logic, Assumptions of Principia Mathematica, Logistics, Proofs of Theorems, etc. "Clearest . . . simplest introduction . . . the intelligent non-mathematician should have no difficulty," MATHEMATICS GAZETTE. Revised, expanded 2nd edition. Truth-value tables. 368pp. 5⅜ 8. S164 Paperbound **$1.75**

TRIGONOMETRICAL SERIES, Antoni Zygmund. On modern advanced level. Contains carefully organized analyses of trigonometric, orthogonal, Fourier systems of functions, with clear adequate descriptions of summability of Fourier series, proximation theory, conjugate series, convergence, divergence of Fourier series. Especially valuable for Russian, Eastern European coverage. 329pp. 5⅜ x 8. S290 Paperbound **$1.50**

THE LAWS OF THOUGHT, George Boole. This book founded symbolic logic some 100 years ago. It is the 1st significant attempt to apply logic to all aspects of human endeavour. Partial contents: derivation of laws, signs and laws, interpretations, eliminations, conditions of a perfect method, analysis, Aristotelian logic, probability, and similar topics. xvii + 424pp. 5⅜ x 8. S28 Paperbound **$2.00**

SYMBOLIC LOGIC, C. I. Lewis, C. H. Langford. 2nd revised edition of probably most cited book in symbolic logic. Wide coverage of entire field; one of fullest treatments of paradoxes; plus much material not available elsewhere. Basic to volume is distinction between logic of extensions and intensions. Considerable emphasis on converse substitution, while matrix system presents supposition of variety of non-Aristotelian logics. Especially valuable sections on strict limitations, existence theorems. Partial contents: Boole-Schroeder algebra; truth value systems, the matrix method; implication and deductibility; general theory of propositions; etc. "Most valuable," Times, London. 506pp. 5⅜ x 8. S170 Paperbound **$2.00**

GROUP THEORY AND LINEAR ALGEBRA, SETS, ETC.

LECTURES ON THE ICOSAHEDRON AND THE SOLUTION OF EQUATIONS OF THE FIFTH DEGREE, Felix Klein. Solution of quintics in terms of rotations of regular icosahedron around its axes of symmetry. A classic, indispensable source for those interested in higher algebra, geometry, crystallography. Considerable explanatory material included. 230 footnotes, mostly bibliography. "Classical monograph . . . detailed, readable book," Math. Gazette. 2nd edition. xvi + 289pp. 5⅜ x 8. S314 Paperbound **$1.85**

INTRODUCTION TO THE THEORY OF GROUPS OF FINITE ORDER, R. Carmichael. Examines fundamental theorems and their applications. Beginning with sets, systems, permutations, etc., progresses in easy stages through important types of groups: Abelian, prime power, permutation, etc. Except 1 chapter where matrices are desirable, no higher math is needed. 783 exercises, problems. xvi + 447pp. 5⅜ x 8. S299 Clothbound **$3.95**
S300 Paperbound **$2.00**

THEORY OF GROUPS OF FINITE ORDER, W. Burnside. First published some 40 years ago, still one of clearest introductions. Partial contents: permutations, groups independent of representation, composition series of a group, isomorphism of a group with itself, Abelian groups, prime power groups, permutation groups, invariants of groups of linear substitution, graphical representation, etc. "Clear and detailed discussion . . . numerous problems which are instructive," Design News. xxiv + 512pp. 5⅜ x 8. S38 Paperbound **$2.45**

COMPUTATIONAL METHODS OF LINEAR ALGEBRA, V. N. Faddeeva, translated by C. D. Benster. 1st English translation of unique, valuable work, only one in English presenting systematic exposition of most important methods of linear algebra—classical, contemporary. Details of deriving numerical solutions of problems in mathematical physics. Theory and practice. Includes survey of necessary background, most important methods of solution, for exact, iterative groups. One of most valuable features is 23 tables, triple checked for accuracy, unavailable elsewhere. Translator's note. x + 252pp. 5⅜ x 8. S424 Paperbound **$1.95**

THE CONTINUUM AND OTHER TYPES OF SERIAL ORDER, E. V. Huntington. This famous book gives a systematic elementary account of the modern theory of the continuum as a type of serial order. Based on the Cantor-Dedekind ordinal theory, which requires no technical knowledge of higher mathematics, it offers an easily followed analysis of ordered classes, discrete and dense series, continuous series, Cantor's transfinite numbers. "Admirable introduction to the rigorous theory of the continuum . . . reading easy," Science Progress. 2nd edition. viii + 82pp. 5⅜ x 8. S129 Clothbound **$2.75**
S130 Paperbound **$1.00**

THEORY OF SETS, E. Kamke. Clearest, amplest introduction in English, well suited for independent study. Subdivisions of main theory, such as theory of sets of points, are discussed, but emphasis is on general theory. Partial contents: rudiments of set theory, arbitrary sets, their cardinal numbers, ordered sets, their order types, well-ordered sets, their cardinal numbers. vii + 144pp. 5⅜ x 8. S141 Paperbound **$1.35**

CONTRIBUTIONS TO THE FOUNDING OF THE THEORY OF TRANSFINITE NUMBERS, Georg Cantor. These papers founded a new branch of mathematics. The famous articles of 1895-7 are translated, with an 82-page introduction by P. E. B. Jourdain dealing with Cantor, the background of his discoveries, their results, future possibiilties. ix + 211pp. 5⅜ x 8. S45 Paperbound **$1.25**